T0189928

Springer Proceedings in Mathematics & Statistics

Volume 197

Springer Proceedings in Mathematics & Statistics

This book series features volumes composed of selected contributions from workshops and conferences in all areas of current research in mathematics and statistics, including operation research and optimization. In addition to an overall evaluation of the interest, scientific quality, and timeliness of each proposal at the hands of the publisher, individual contributions are all refereed to the high quality standards of leading journals in the field. Thus, this series provides the research community with well-edited, authoritative reports on developments in the most exciting areas of mathematical and statistical research today.

More information about this series at http://www.springer.com/series/10533

Valery A. Kalyagin · Alexey I. Nikolaev
Panos M. Pardalos · Oleg A. Prokopyev
Editors

Models, Algorithms, and Technologies for Network Analysis

NET 2016, Nizhny Novgorod, Russia, May 2016

Editors
Valery A. Kalyagin
National Research University Higher School of Economics
Nizhny Novgorod
Russia

Alexey I. Nikolaev
National Research University Higher School of Economics
Nizhny Novgorod
Russia

Panos M. Pardalos
Department of Industrial and Systems Engineering
University of Florida
Gainesville, FL
USA

Oleg A. Prokopyev
Department of Industrial Engineering
University of Pittsburgh
Pittsburgh, PA
USA

ISSN 2194-1009 ISSN 2194-1017 (electronic)
Springer Proceedings in Mathematics & Statistics
ISBN 978-3-319-86012-1 ISBN 978-3-319-56829-4 (eBook)
DOI 10.1007/978-3-319-56829-4

Mathematics Subject Classification (2010): 90-02, 90C31, 90C27, 90C09, 90C10, 90C11, 90C35, 90B06, 90B10, 90B18, 90B40, 68R01

Printed on acid-free paper

This Springer imprint is published by Springer Nature
The registered company is Springer International Publishing AG
The registered company address is: Gewerbestrasse 11, 6330 Cham, Switzerland

Preface

This volume is based on the papers presented at the 6th International Conference on Network Analysis held in Nizhny Novgorod, Russia, May 26–28, 2016. The main focus of the conference and this volume is centered around the development of new computationally efficient algorithms as well as underlying analysis and optimization of graph (network) structures induced either by natural or by artificial complex networks. Various applications to social networks, power transmission grids, stock market networks, and human brain networks are also considered. The previous books based on the papers presented at the 1st–5th Conferences International Conference on Network Analysis can be found in [1–5]. The current volume consists of three major parts, namely optimization approaches, network models, and related applications, which we briefly overview next.

Part I of this book is focused on optimization problems in networks. In Chapter "Linear Max-Min Fairness in Multi-commodity Flow Networks," a linear max-min fairness (LMMF) approach using goal programming is proposed. This model can be applied to max-min fairness (MMF) problems in networks with applications to multicommodity flows in networks. The proposed model provides a high flexibility for the decision maker to determine the level of throughput and the fairness of flow in the network.

In Chapter "Heuristic for Maximizing Grouping Efficiency in the Cell Formation Problem," Cell Formation Problem in Group Technology with grouping efficiency as an objective function is considered. A heuristic approach for obtaining high-quality solutions is presented. The computational results show the effectiveness of the approach.

In Chapter "Efficient Methods of Multicriterial Optimization Based on the Intensive Use of Search Information," an efficient approach for solving complex multicriterial optimization problems is developed. In particular, it is based on reducing multicriterial problems to nonlinear programming problems via the minimax convolution of the partial criteria, reducing dimensionality by using Peano evolvents, and applying efficient information-statistical global optimization methods. The results of the computational experiments show that the proposed approach reduces the computational costs of solving multicriterial optimization problems.

In Chapter "Comparison of Two Heuristic Algorithms for a Location and Design Problem," the special case of the location and design problem is considered. A Variable Neighborhood Search algorithm and a Greedy Weight Heuristic are proposed. In particular, new best known solutions have been found by applying the proposed approaches.

In Chapter "A Class of Smooth Modification of Space-Filling Curves for Global Optimization Problems," a class of smooth modification of space-filling curves applied to global optimization problems is presented. These modifications make the approximations of the Peano curves (evolvents) differentiable in all points and save the differentiability of the optimized function. Some results of numerical experiments with the original and modified evolvents for solving global optimization problems are discussed.

In Chapter "Iterative Local Search Heuristic for Truck and Trailer Routing Problem," Site-Dependent Truck and Trailer Routing Problem with Hard and Soft Time Windows and Split Deliveries is considered. A new iterative local search heuristic for solving this problem was developed.

Part II of this book presents several network models. Chapter "Power in Network Structures" considers an application of power indices, which take into account the preferences of agents for coalition formation proposed for an analysis of power distribution in elected bodies to reveal most powerful (central) nodes in networks. These indices take into account the parameters of the nodes in networks, a possibility of group influence from the subset of nodes to single nodes, and intensity of short and long interactions among the nodes.

In Chapter "Do Logarithmic Proximity Measures Outperform Plain Ones in Graph Clustering?," a number of graph kernels and proximity measures as well as the corresponding distances were applied for clustering nodes in random graphs and several well-known datasets. In the experiments, the best results are obtained for the logarithmic Communicability measure. However, some more complicated cases are indicated in which other measures, typically Communicability and plain Walk, can be the winners.

In Chapter "Analysis of Russian Power Transmission Grid Structure: Small World Phenomena Detection," the spatial and topological structure of the Unified National Electricity Grid (UNEG)—Russia's power transmission grid—is analyzed. The research is focused on the applicability of the small-world model to the UNEG network. For this purpose, geo-latticization algorithm has been developed. As a result of applying the new method, a reliable conclusion has been made that the small-world model is applicable to the UNEG.

In Chapter "A New Approach to Network Decomposition Problems," a new approach to network decomposition problems is suggested. The suggested approach is focused on construction of a family of classifications. Based on this family, two numerical indices are introduced and calculated. This approach was applicable to political voting body and stock market.

In Chapter "Homogeneity Hypothesis Testing for Degree Distribution in the Market Graph," the problem of homogeneity hypothesis testing for degree distribution in the market graph is studied. Multiple hypotheses testing procedure is proposed and applied for China and India stock markets. The procedure is constructed using bootstrap method for individual hypotheses and Bonferroni correction for multiple testing.

Chapter "Stability Testing of Stock Returns Connections" considers the testing problem of connection stability which is formulated as homogeneity hypothesis testing of several covariance matrices for multivariate normal distributions of stock returns. New procedure is proposed and applied for stability testing of connections for French and German stock markets.

Part III of this book is focused on applications of network analysis. In Chapter "Network Analysis of International Migration," the network approach to the problem of international migration is employed. The international migration can be represented as a network where the nodes correspond to countries and the edges correspond to migration flows. The main focus of the study is to reveal a set of critical or central elements in the network.

In Chapter "Overlapping Community Detection in Social Networks with Node Attributes by Neighborhood Influence," a fast method for overlapping community detection in social networks with node attributes is presented. The proposed algorithm is based on attribute transfer from neighbor vertices and does not require any knowledge of attributes meaning. Computational results show that the proposed method outperforms other algorithms such as Infomap, modularity maximization, CESNA, BigCLAM, and AGM-fit.

In Chapter "Testing Hypothesis on Degree Distribution in the Market Graph," the problem of testing hypotheses on degree distribution in the market graph is discussed. Research methodology of power law hypothesis testing is presented. This methodology is applied to testing hypotheses on degree distribution in the market graphs for different stock markets.

In Chapter "Application of Network Analysis for FMCG Distribution Channels," the approach for multidimensional analysis of marketing tactics of the companies employing network tools is presented. The research suggests omni-channel distribution tactic of a company as a node in eight-dimensional space. Empirical implication is approved on the sample from 5694 distributors from sixteen fast-moving consumer goods-distributing companies from direct selling industry.

In Chapter "Machine Learning Application to Human Brain Network Studies: A Kernel Approach," a task of predicting normal and pathological phenotypes from macroscale human brain networks is considered. The research focuses on kernel classification methods. It presents the results of performance comparison of the different kernels in tasks of classifying autism spectrum disorder versus typical development and carriers versus non-carriers of an allele associated with an increased risk of Alzheimer's disease.

In Chapter "Co-author Recommender System," a new recommender system for finding possible collaborator with respect to research interests is proposed. The recommendation problem is formulated as a link prediction within the co-authorship network. The network is derived from the bibliographic database and enriched by the information on research papers obtained from Scopus and other publication ranking systems.

Chapter "Network Studies in Russia: From Articles to the Structure of a Research Community" focuses on the structure of a research community of Russian scientists involved in network studies by analysis of articles published in Russian-language journals. It covers the description of method of citation (reference) analysis that is used and the process of data collection from eLibrary.ru resource, as well as presents some brief overview of collected data (based on analysis of 8000 papers).

We would like to take this opportunity to thank all the authors and referees for their efforts. This work is supported by the Laboratory of Algorithms and Technologies for Network Analysis (LATNA) of the National Research University Higher School of Economics.

Nizhny Novgorod, Russia Valery A. Kalyagin
Nizhny Novgorod, Russia Alexey I. Nikolaev
Gainesville, FL, USA Panos M. Pardalos
Pittsburgh, PA, USA Oleg A. Prokopyev

References

1. Goldengorin, B.I., Kalyagin, V.A., Pardalos, P.M. (eds.): Models, algorithms and technologies for network analysis. In: Proceedings of the First International Conference on Network Analysis. Springer Proceedings in Mathematics and Statistics, vol. 32. Springer, Cham (2013a)
2. Goldengorin, B.I., Kalyagin, V.A., Pardalos, P.M. (eds.): Models, algorithms and technologies for network analysis. In: Proceedings of the Second International Conference on Network Analysis. Springer Proceedings in Mathematics and Statistics, vol. 59. Springer, Cham (2013b)
3. Batsyn, M.V., Kalyagin, V.A., Pardalos, P.M. (eds.): Models, algorithms and technologies for network analysis. In: Proceedings of Third International Conference on Network Analysis. Springer Proceedings in Mathematics and Statistics, vol. 104. Springer, Cham (2014)
4. Kalyagin, V.A., Pardalos, P.M., Rassias, T.M. (eds.): Network Models in Economics and Finance. Springer Optimization and Its Applications, vol. 100. Springer, Cham (2014)
5. Kalyagin, V.A., Koldanov, P.A., Pardalos, P.M. (eds.): Models, algorithms and technologies for network analysis. In: NET 2014, Nizhny Novgorod, Russia, May 2014. Springer Proceedings in Mathematics and Statistics, vol. 156. Springer, Cham (2016)

Contents

Contributors

Fuad Aleskerov National Research University Higher School of Economics (HSE), Moscow, Russia; V.A. Trapeznikov Institute of Control Sciences of Russian Academy of Sciences (ICS RAS), Moscow, Russia

Mikhail Batsyn Laboratory of Algorithms and Technologies for Network Analysis, National Research University Higher School of Economics, Nizhny Novgorod, Russia

Hamoud Bin Obaid University of Oklahoma, Norman, OK, USA; King Saud University, Riyadh, Saudi Arabia

Oleg Bulanov National Research University Higher School of Economics, Moscow, Russia

Ilya Bychkov Laboratory of Algorithms and Technologies for Network Analysis, National Research University Higher School of Economics, Nizhny Novgorod, Russia

Pavel Chebotarev Institute of Control Sciences of the Russian Academy of Sciences, Moscow, Russia; The Kotel'nikov Institute of Radio-engineering and Electronics (IRE) of Russian Academy of Sciences, Moscow, Russia

Vladislav Chesnokov Bauman Moscow State Technical University, Moscow, Russia

Yulia Dodonova National Research University Higher School of Economics, Moscow, Russia

Victor Gergel Nizhni Novgorod State University, Nizhny Novgorod, Russia

Alexander Gnusarev Sobolev Institute of Mathematics SB RAS, Omsk Branch, Omsk, Russia

Alexey Goryachih Lobachevsky State University of Nizhni Novgorod, Nizhni Novgorod, Russia

Ivan S. Grechikhin National Research University Higher School of Economics, Laboratory of Algorithms and Technologies for Network Analysis, Nizhny Novgorod, Russia

Vladimir Ivashkin Moscow Institute of Physics and Technology, Dolgoprudny, Moscow Region, Russia

Ilia Karpov International Laboratory for Applied Network Research, National Research University Higher School of Economics, Moscow, Russia

P.A. Koldanov Laboratory of Algorithms and Technologies for Network Analysis, National Research University Higher School of Economics, Nizhny Novgorod, Russia

Nadezda Kolesnik National Research University Higher School of Economics, Moscow, Russia

Evgeny Kozinov Nizhni Novgorod State University, Nizhny Novgorod, Russia

Anvar Kurmukov National Research University Higher School of Economics, Moscow, Russia

Valentina Kuskova National Research University Higher School of Economics, Moscow, Russia

J.D. Larushina National Research University Higher School of Economics, Laboratory of Algorithms and Technologies for Network Analysis, Nizhny Novgorod, Russia

Ilya Makarov National Research University Higher School of Economics, Moscow, Russia

Sergey Makrushin Financial University Under the Government of the Russian Federation, Moscow, Russia

Daria Maltseva International Laboratory for Applied Network Research, National Research University Higher School of Economics, Moscow, Russia

Natalia Meshcheryakova National Research University Higher School of Economics (HSE), Moscow, Russia; V.A. Trapeznikov Institute of Control Sciences of Russian Academy of Sciences (ICS RAS), Moscow, Russia

Panos M. Pardalos Laboratory of Algorithms and Technologies for Network Analysis, National Research University Higher School of Economics, Nizhny Novgorod, Russia; Center for Applied Optimization, University of Florida, Gainesville, USA

Anna Rezyapova National Research University Higher School of Economics (HSE), Moscow, Russia

Alexander Rubchinsky National Research University "Higher School of Economics", Moscow, Russia; National Research Technological University "MISIS", Moscow, Russia

D.P. Semenov Laboratory of Algorithms and Technologies for Network Analysis, National Research University Higher School of Economics, Nizhny Novgorod, Russia

Sergey Shvydun National Research University Higher School of Economics (HSE), Moscow, Russia; V.A. Trapeznikov Institute of Control Sciences of Russian Academy of Sciences (ICS RAS), Moscow, Russia

Theodore B. Trafalis University of Oklahoma, Norman, OK, USA

Olga Tretyak National Research University Higher School of Economics, Moscow, Russia

M.A. Voronina Laboratory of Algorithms and Technologies for Network Analysis, Nizhny Novgorod, National Research University Higher School of Economics, Moscow, Russia

Leonid E. Zhukov National Research University Higher School of Economics, Moscow, Russia

Part I
Optimization

Linear Max-Min Fairness in Multi-commodity Flow Networks

Hamoud Bin Obaid and Theodore B. Trafalis

Abstract In this paper, a linear max-min fairness (LMMF) approach using goal programming is proposed. The linear model in this approach is a bi-objective model where the flow is maximized in one objective, and the fairness in flow is maximized for the other objective. This model can be applied to max- min fairness (MMF) problems in networks with applications to multi-commodity flows in networks. The proposed model provides high flexibility for the decision maker to determine the level of throughput and the fairness of flow in the network. The goal of this paper is to find a less-complex approach to find MMF flow in multi-commodity networks. An example is presented to illustrate the methodology.

1 Introduction

The basic approach to apply MMF [1] on a problem is to start with the lowest capacity object of a system to fill with the available resource to its maximum capacity, then fill the second lowest object to its capacity continuing with the next lowest object until the available resource ends. MMF insures that the resource is fairly distributed among the available objects starting with the lower capacity ones. MMF is widely applied in traffic engineering and load-balancing problems. Internet protocol (IP) is an extensively studied application using MMF, where the goal is to maximize the throughput and insure fair distribution of resources. A new approach has not been proposed, according to our knowledge, to maximize the throughput of a flow network and maximize its fairness. A bi-objective model is implemented to optimize the two objectives in a flow network. The proposed approach to solve the bi-objective model is to use the ε-constraint method. In the next section, a brief MMF overview from

H. Bin Obaid (✉) · T.B. Trafalis
University of Oklahoma, 202 W. Boyd St., Lab 28, Norman, OK 73019, USA
e-mail: hsbinobaid@ou.edu; hsbinobaid@ksu.edu.sa

T.B. Trafalis
e-mail: ttrafalis@ou.edu

H. Bin Obaid
King Saud University, P.O. BOX 800, Riyadh 11421, Saudi Arabia

V.A. Kalyagin et al. (eds.), *Models, Algorithms, and Technologies for Network Analysis*, Springer Proceedings in Mathematics & Statistics 197,
DOI 10.1007/978-3-319-56829-4_1

related work is presented. In Sect. 2 the solution approach is discussed. In Sect. 3 a toy example is presented and the results of the model are discussed with a conclusion in Sect. 3.

1.1 MMF in Networks

The early development of the MMF approach in networks appeared in 1974 and is based on lexicographic maximization [2]. MMF is commonly applied to multi-commodity flow networks and is used for internet protocol (IP) throughput optimization. There has been a number of research papers in this field, such as [3–6]. Although the flow in IP network is unsplittable, Multi-Protocol Label Switching (MPLS) technology allows the flow in IP to be splitted among different paths. This technology enables us to design a more relaxed model by not forcing the flow to be routed through one path, which reduces the complexity of solving this problem. The most common approach to apply MMF on multi-commodity flow networks is to solve an LP model a number of times depending on the size of the network [1, 7, 8]. The constraints are classified into two sets: non-blocking constraints and blocking constraints. The blocking constraints set is initialized to be empty. Then an LP is solved until a blocking constraint is found, then the constraint is moved from the non-blocking constraints set to the blocking set until an empty set of non-blocking constraints is reached. This process is computationally demanding for very large networks due to the approach of identifying the blocking constraints in each iteration. The approach to identify the blocking constraints is by using the strict complementary slackness. The constraint is blocking if it is binding which leads to a zero slack value. A positive value of the corresponding dual variable is used as an indicator of a blocking constraint. However, this condition is unnecessary but insures convergence. It indicates that not all blocking constraints in one iteration can be identified, and we call that degeneracy. This leads to a higher number of iterations to find the MMF flow in the network. In [7], the authors proposed another approach called binary search that reduces the number of iterations to identify the blocking constraints. Another method to find MMF in [9] is by the polyhedral approach. Geometry is used to find MMF in networks, where the number of commodities represents the number of dimensions of the polyhedron. Changing the flow of one commodity would result in change of flow for other commodities sharing the same capacity of the network forming a polyhedron. The MMF flow point in the polyhedron can be located by maximizing the distance between the point and the zero axes for each commodity. The drawbacks of this approach is that the routing path for each commodity has to be predefined to identify the right-hand side values of the constraints, and the model is solved iteratively which can lead to high computational time. For the polyhedral approach, there is no efficient model that has been developed yet to find the MMF flow on the network.

In this paper, the overall number of iterations is considerably reduced leading to an efficient approach to find MMF flow for larger networks.

2 Solution Methodology

Given the network N, described through the graph G = (V, E) with the set of nodes V and set of directed arcs E. Each arc e ∈ E, where e = (i, j) and i, j ∈ V. s, t ∈ V where s is a supply node and t is a terminal node. S is the supply node set, and T is the terminal node set where s ∈ S and t ∈ to T. The model is composed of hard and soft constraints since a goal programming approach is used. The set of constraints (1) are known to be the flow conservation constraint, where x_{ij}^k is the commodity flow k from node i to node j. If the incoming flow is greater than the outgoing flow at node i, then node i is a source node. If the outgoing flow is greater than the incoming flow at node i, then node i is a source node. If the incoming and outgoing flows are equal at node i, the node i is a transshipment node.

$$\sum_{j \in N} x_{ij}^k - x_{j,i}^k = \begin{cases} f_i^k & \text{if } i = s, \\ -f_i^k & \text{if } i = t, \\ 0 & \text{otherwise.} \end{cases} \tag{1}$$

With the additional index k, each commodity flows is in a distinct network. However, the set of capacity constraints (2) link all the commodity flows to one network to share the same capacity resource. It insures that all the commodity flows pass through the arc e = (i, j) but do not exceed the capacity of the arc.

$$\sum_{k \in K} x_{ij}^k \leq C_{ij} \qquad \forall (i, j) \in E, k \in K \tag{2}$$

The set of soft constraints (3) provide a way to reduce the difference among the commodity flows resulting in fair allocation of commodity flows. d^{kl} is the deviational variable of commodity flow k compared with commodity flow l. These deviational variables make up the difference between two commodity flows.

$$\sum_{i \in S} f_i^k - \sum_{j \in S} f_j^l + d^{kl} - d^{lk} = 0 \qquad \forall k \in K \tag{3}$$

If the deviational variables are minimized to 0, all the commodity flows are equal and said to be fairly distributed. The sum of commodity flows $\sum_{i \in S} f_i^k$ appears in the case of having multiple source nodes. The set of constraints (4) are the nonnegativity constraints.

$$x_{ij}^k, f_i^k, d^{kl} \geq 0 \qquad \forall (i, j) \in E, i \in V, k \in K \tag{4}$$

The first objective (5) is to maximize the overall flow to utilize the available capacity resources. Maximizing the first objective does not lead to a fair distribution of flow. However, maximizing the flow as a first step is useful to adjust the value of ε when the first objective is set as a constraint. When the first objective is set as a constraint using the ε-constraint method, the sum of deviations is minimized in the second objective (6).

$$Max\ Z_1 = \sum_{i \in N, k \in K} f_i^k \tag{5}$$

$$Min\ Z_2 = \sum_{k,l \in K} d^{kl} \tag{6}$$

The decision maker then decides what level of fairness is desired considering the tradeoff between fairness and total flow.

2.1 MMF in Networks

Let us consider the simple example of a network (N1) in [1] where we have three routers A, B, and C and three links AB, BC, and AC with capacities 2 MB/s, 3 MB/s, and 1 MB/s respectively. If the flow is maximized, the resulting flow is illustrated in Fig. 1a. The total throughput is 6.

The value of the maximum flow is determined. The next step is to set the first objective (5) as a constraint using the ε-constraint method.

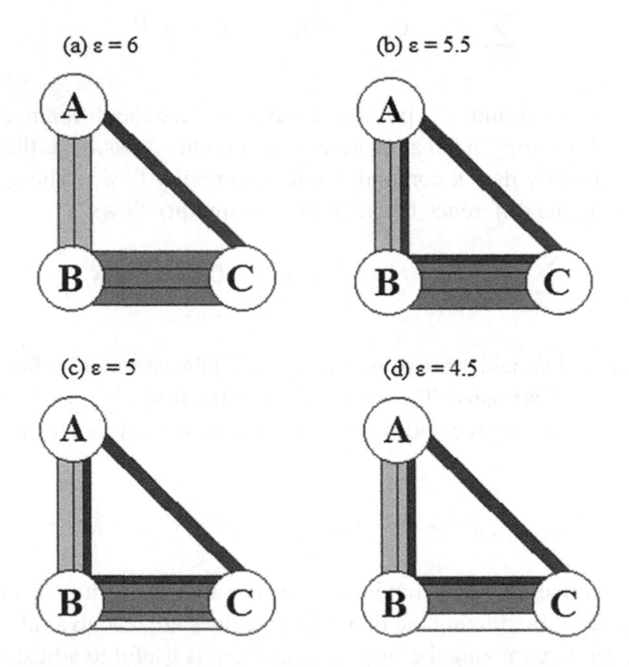

Fig. 1 The result when maximizing the overall throughput. When maximizing the overall flow, the deviation is maximum resulting in unfair distributution of commodity flows as seen in (**a**). If the overall flow is reduced to 5.5, the deviation is reduced resulting in MMF solution in (**b**). If the flow is reduced below 5.5, the network is not utilized as seen in (**c**) and (**d**)

Table 1 Summary of the results using different values of ε

ε (Total flow)	6		5.5		5		4.5	
Connection	Path	Flow	Path	Flow	Path	Flow	Path	Flow
A $\longrightarrow$ B	A-B	2	A-B	1.5	A-B	1.5	A-B	1.5
A $\longrightarrow$ C	A-C	1	A-B-C+A-C	1.5	A-B-C+A-C	1.5	A-B-C+A-C	1.5
B $\longrightarrow$ C	B-C	3	B-C	2.5	B-C	2	B-C	1.5
Min deviation	4		2		1		0	

$$Min\ Z = \sum_{k,l \in K} d^{kl} \tag{7}$$

s.t.

$$\sum_{i \in N, k \in K} f_i^k \geq \varepsilon \tag{8}$$

$$\sum_{j \in N} x_{ij}^k - x_{j,i}^k = \begin{cases} f_i^k & \text{if } i = s, \\ -f_i^k & \text{if } i = t, \\ 0 & \text{otherwise.} \end{cases} \tag{9}$$

$$\sum_{k \in K} x_{ij}^k \leq C_{ij} \qquad \forall (i,j) \in E, k \in K \tag{10}$$

$$\sum_{i \in S} f_i^k - \sum_{j \in S} f_j^l + d^{kl} - d^{lk} = 0 \qquad \forall k \in K \tag{11}$$

$$x_{ij}^k, f_i^k, d^{kl} \geq 0 \qquad \forall (i,j) \in E, i \in V, k \in K \tag{12}$$

When the ε value is set to 6, the maximum flow in this case, the minimum sum of deviations obtained is 4. If ε is reduced to 5.5, the minimum sum of deviations resulted is 2 as shown in Fig. 1b. However, the minimum sum of deviations becomes 1 if we set ε to 5, but the set of capacity constraints (10) are no longer binding, which indicates that the capacity resource is not fully utilized. Table 1 summarizes the results of the tested example. The deviation can be minimized to zero if the ε value is set to 4.5, resulting in equal flows for all commodities with some nonbinding capacity constraints as seen in Fig. 1c, d.

We can observe the results in Table 1, where the sum of deviational variables decreases as the ε value decreases. Reducing the flow reduces the congestion resulted from the competing commodities trying to reach to their destinations resulting in giving space for sharing as seen in the example above. In Fig. 2, there are infinite nondominated solutions creating a Pareto front creating a convex objective space. Additionally, this proposed approach requires less computational time and provides high flexibility in terms of decision-making. This problem is solved in a polynomial time [10].

If the value of ε is continuous, the resulting Pareto front can be observed in Fig. 3. It can be noticed that the slope is different in the intervals [4.5, 5.5) and (5.5, 6]. The most attractive value of ε is 5.5, which is the value we would obtain if the MMF algorithm was used [1]. The reason that 5.5 is the most attractive value is because it gains the most of the two competing objectives. Rationally, if the ε value is decreased below the value 5.5, the gain in fairness is not substantial compared to the gain acquired by creating space for sharing.

The proposed algorithm is applied on a random network with 26 nodes, 83 links, and 109 commodities. The results can be observed in Fig. 4 showing that the level of fairness decreases to 0.1 of the maximum deviation when we reduce the overall flow to 0.6. In Fig. 4, the MMF solution exists with solving the model only several times

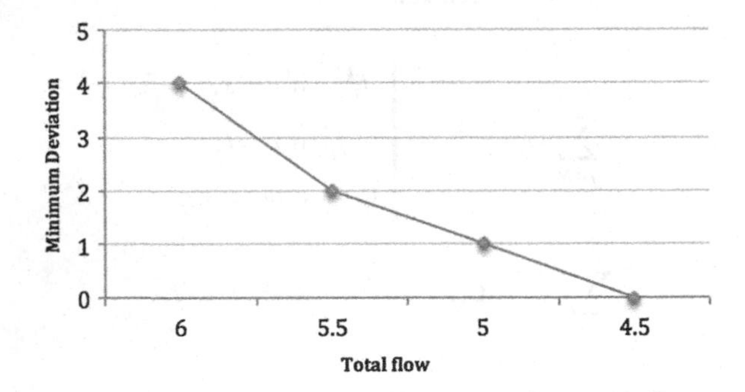

Fig. 2 Pareto front for the 4 values of ε

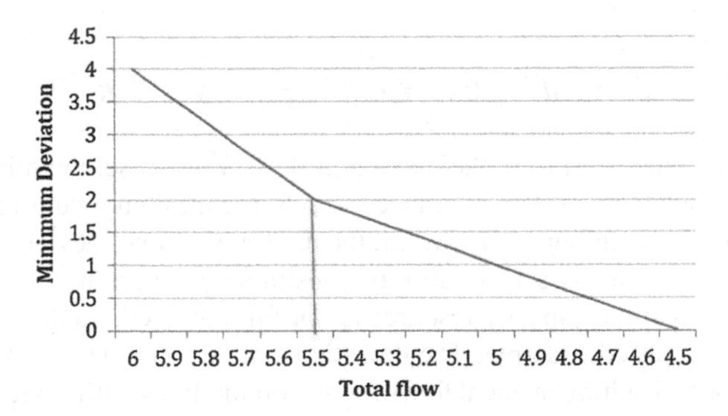

Fig. 3 Pareto front for continuous value of ε

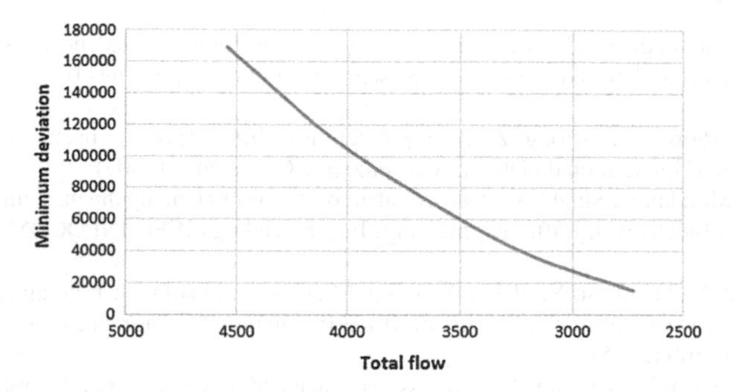

Fig. 4 Pareto front for continuous value of ε

regardless of the size the network. However, in the future version of this model we will propose a robust approach to decide which value of ε gives the MMF solution for any network size.

3 Conclusion

In this paper, a linear max-min fairness (LMMF) approach using goal programming is proposed. The resulting linear model was described as a bi-objective model where we maximize the flow as the first objective, and the fairness in flow as the second objective. A small example from communication networks was used to illustrate the idea. We can conclude that every network structure affects the behavior of the Pareto front of the two objectives. In addition, other performance measures will be included to improve the accuracy when selecting the value of ε. Applying this model to large-scale networks will be the objective of future research. Robust techniques to select the ε value will be also explored. Selection of ε is critical in utilizing the resource capacity of the network. Moreover, another objective is to apply the model on a variety of application such as is congestion control, traffic engineering, and IP networks.

References

1. Nace, D., Pi, M.: Max-min fairness and its applications to routing and load-balancing in communication networks: a tutorial. IEEE Commun. Surv. Tutor. **10**(4), 5–17 (2008)
2. Megiddo, N.: Optimal flows in networks with multiple sources and sinks. Math. Program. **7**, 97–107 (1974)
3. Nace, D., Doan, N., Gourdin, E., Liau, B.: Computing optimal max-min fair resource allocation for elastic flows, EEE Trans. Netw. **14**, 1272–1281 (2006)
4. Bertsekas, D., Hall, P.: Data Networks. Prentice-Hall, Engelwood Cliffs, N.J. (1992)

5. Chen, S., Nahrstedt, K.: Maxmin fair routing in connection-oriented networks 1 introduction. In: Euro-Parallel and Distributed Systems Conference, pp. 163–168, Vienna Austria, (July 1998)
6. Nilsson, P., Pioro, M., Dziong, Z.: Link protection within an existing backbone network. In: Proceedings of International Network Optimization Conference (2003)
7. Danna, E., Mandal, S., Singh, A.: A practical algorithm for balancing the max-min fairness and throughput objectives in traffic engineering. In: Proceedings IEEE INFOCOM, pp. 846–854 (2012)
8. Mollah, Md.A., YuanFast, X., Pakin, S., Lang, M.: Calculation of max-min fair rates for multi-commodity flows in fat-tree networks. In: IEEE International Conference on Cluster Computing, pp. 351–360 (2015)
9. Retvari, G., Biro, J.J., Cinkler, T.: Fairness in capacitated networks: a polyhedral approach. In: IEEE INFOCOM 2007-26th IEEE International Conference on Computer Communications, pp. 1604–1612 (2007)
10. Cohen, E., Megiddo, N.: Algorithms and complexity analysis for some flow problems. Algorithmica **11**(3), 320–340 (1994). (March 1994)

Heuristic for Maximizing Grouping Efficiency in the Cell Formation Problem

Ilya Bychkov, Mikhail Batsyn and Panos M. Pardalos

Abstract In our paper, we consider the Cell Formation Problem in Group Technology with grouping efficiency as an objective function. We present a heuristic approach for obtaining high-quality solutions of the CFP. The suggested heuristic applies an improvement procedure to obtain solutions with high grouping efficiency. This procedure is repeated many times for randomly generated cell configurations. Our computational experiments are performed for popular benchmark instances taken from the literature with sizes from 10×20 to 50×150. Better solutions unknown before are found for 23 instances of the 24 considered. The preliminary results for this paper are available in Bychkov et al. (Models, algorithms, and technologies for network analysis, Springer, NY, vol. 59, pp. 43–69, 2013, [7]).

1 Introduction

Flanders [15] was the first who formulated the main ideas of the group technology. The notion of the Group Technology was introduced in Russia by [30], though his work was translated to English only in 1966 [31]. One of the main problems stated by the Group Technology is the optimal formation of manufacturing cells, i.e., grouping of machines and parts into cells such that for every machine in a cell the number of the parts from this cell processed by this machine is maximized and the number of the parts from other cells processed by this machine is minimized. In other words,

I. Bychkov (✉) · M. Batsyn · P.M. Pardalos
Laboratory of Algorithms and Technologies for Network Analysis,
National Research University Higher School of Economics,
136 Rodionova Street, Nizhny Novgorod 603093, Russia
e-mail: ibychkov@hse.ru

M. Batsyn
e-mail: mbatsyn@hse.ru

P.M. Pardalos
Center for Applied Optimization, University of Florida, 401 Weil Hall,
P.O. Box 116595, Gainesville 32611-6595, USA
e-mail: pardalos@ufl.edu

V.A. Kalyagin et al. (eds.), *Models, Algorithms, and Technologies for Network Analysis*, Springer Proceedings in Mathematics & Statistics 197,
DOI 10.1007/978-3-319-56829-4_2

the intra-cell loading of machines is maximized and simultaneously the inter-cell movement of parts is minimized. This problem is called the Cell Formation Problem (CFP). Burbidge [5] suggested his Product Flow Analysis (PFA) approach for the CFP, and later popularized the Group Technology and the CFP in his book [6].

The CFP is NP-hard since it can be reduced to the clustering problem [16]. That is why there is a great number of heuristic approaches for solving CFP and almost no exact ones. The first algorithms for solving the CFP were different clustering techniques. Array-based clustering methods find rows and columns permutations of the machine-part matrix in order to form a block-diagonal structure. These methods include: Bond Energy Algorithm (BEA) of [29], Rank Order Clustering (ROC) algorithm by [20], its improved version ROC2 by [21], Direct Clustering Algorithm (DCA) of [12], Modified Rank Order Clustering (MODROC) algorithm by [9], the Close Neighbor Algorithm (CAN) by [4]. Hierarchical clustering methods at first form several big cells, then divide each cell into smaller ones and so on gradually improving the value of the objective function. The most well-known methods are Single Linkage [28], Average Linkage [39], and Complete Linkage [32] algorithms. Nonhierarchical clustering methods are iterative approaches which start from some initial partition and improve it iteratively. The two most successful are GRAFICS algorithm by [41] and ZODIAC algorithm by [10]. A number of works considered the CFP as a graph partitioning problem, where machines are vertices of a graph. [37] used clique partitioning of the machines graph. Askin and Chiu [2] implemented a heuristic partitioning algorithm to solve CFP. Ng [35, 36] suggested an algorithm based on the minimum spanning tree problem. Mathematical programming approaches are also very popular for the CFP. Since the objective function of the CFP is rather complicated from the mathematical programming point of view most of the researchers use some approximation model which is then solved exactly for small instances and heuristically for large. [25] formulated CFP via p-median model and solved several small size CFP instances, [40] used Generalized Assignment Problem as an approximation model, [44] proposed a simplified p-median model for solving large CFP instances, [22] applied minimum k-cut problem to the CFP, [17] used p-median approximation model and solved it exactly by means of their pseudo-boolean approach including large CFP instances up to 50×150 instance. A number of meta-heuristics have been applied recently to the CFP. Most of these approaches can be related to genetic, simulated annealing, Tabu search, and neural networks algorithms. Among them are works such as: [18, 26, 27, 45–47].

Our heuristic algorithm is based on sequential improvements of the solution. We modify the cell configuration by enlarging one cell and reducing another. The basic procedure of the algorithm has the following steps:

1. Generate a random cell configuration.
2. Improve the initial solution moving one row or column from one cell to another until the grouping efficiency is increasing.
3. Repeat steps 1–2 a predefined number of times (we use 2000 times for computational experiments in this paper).

The paper is organized as follows. In the next section, we provide the Cell Formation Problem formulation. In Sect. 3 we present our improvement heuristic that allows us to get good solutions by iterative modifications of cells which lead to increasing of the objective function. In Sect. 4 we report our computational results and Sect. 5 concludes the paper with a short summary.

2 The Cell Formation Problem

The CFP consists in an optimal grouping of the given machines and parts into cells. The input for this problem is given by m machines, p parts, and a rectangular machine-part incidence matrix $A = [a_{ij}]$, where $a_{ij} = 1$ if part j is processed on machine i. The objective is to find an optimal number and configuration of rectangular cells (diagonal blocks in the machine-part matrix) and optimal grouping of rows (machines) and columns (parts) into these cells such that the number of zeros inside the chosen cells (voids) and the number of ones outside these cells (exceptions) are minimized. A concrete combination of rectangular cells in a solution (diagonal blocks in the machine-part matrix) we will call a cells configuration. Since it is usually not possible to minimize these two values simultaneously there have appeared a number of compound criteria trying to join it into one objective function. Some of them are presented below.

For example, we are given the machine-part matrix [43] shown in Table 1. Two different solutions for this CFP are shown in Tables 2 and 3. The left solution is better because it has less voids (3 against 4) and exceptions (4 against 5) than the right one. But one of its cells is a singleton—a cell which has less than two machines or parts.

Table 1 Machine-part 5 × 7 matrix from [43]

	p_1	p_2	p_3	p_4	p_5	p_6	p_7
m_1	1	0	0	0	1	1	1
m_2	0	1	1	1	1	0	0
m_3	0	0	1	1	1	1	0
m_4	1	1	1	1	0	0	0
m_5	0	1	0	1	1	1	0

Table 2 Solution with singletons

	p_7	p_6	p_1	p_5	p_3	p_2	p_4
m_1	1	1	1	1	0	0	0
m_4	0	0	1	0	1	1	1
m_3	0	1	0	1	1	0	1
m_2	0	0	0	1	1	1	1
m_5	0	1	0	1	0	1	1

Table 3 Solution without singletons

	p_7	p_1	p_6	p_5	p_4	p_3	p_2
m_1	1	1	1	1	0	0	0
m_4	0	1	0	0	1	1	1
m_2	0	0	0	1	1	1	1
m_3	0	0	1	1	1	1	0
m_5	0	0	1	1	1	0	1

In some CFP formulations singletons are not allowed, so in this case this solution is not feasible. In this paper, we consider both the cases (with allowed singletons and with not allowed) and when there is a solution with singletons found by the suggested heuristic better than without singletons we present both the solutions.

There are a number of different objective functions used for the CFP. The following four functions are the most widely used:

1. Grouping efficiency suggested by [11]:

$$\eta = q\eta_1 + (1-q)\eta_2, \tag{1}$$

where

$$\eta_1 = \frac{n_1 - n_1^{out}}{n_1 - n_1^{out} + n_0^{in}} = \frac{n_1^{in}}{n^{in}},$$

$$\eta_2 = \frac{mp - n_1 - n_0^{in}}{mp - n_1 - n_0^{in} + n_1^{out}} = \frac{n_0^{out}}{n^{out}},$$

η_1—a ratio showing the intra-cell loading of machines (or the ratio of the number of ones in cells to the total number of elements in cells).
η_2—a ratio inverse to the inter-cell movement of parts (or the ratio of the number of zeroes out of cells to the total number of elements out of cells).
q—a coefficient ($0 \le q \le 1$) reflecting the weights of the machine loading and the inter-cell movement in the objective function. It is usually taken equal to $\frac{1}{2}$, which means that it is equally important to maximize the machine loading and minimize the inter-cell movement.
n_1—a number of ones in the machine-part matrix,
n_0—a number of zeroes in the machine-part matrix,
n^{in}—a number of elements inside the cells,
n^{out}—a number of elements outside the cells,

n_1^{in}—a number of ones inside the cells,
n_1^{out}—a number of ones outside the cells,
n_0^{in}—a number of zeroes inside the cells,
n_0^{out}—a number of zeroes outside the cells.

2. Grouping efficacy suggested by [23]:

$$\tau = \frac{n_1 - n_1^{out}}{n_1 + n_0^{in}} = \frac{n_1^{in}}{n_1 + n_0^{in}} \tag{2}$$

3. Group Capability Index (GCI) suggested by [19]:

$$GCI = 1 - \frac{n_1^{out}}{n_1} = \frac{n_1 - n_1^{out}}{n_1} \tag{3}$$

4. Number of exceptions (ones outside cells) and voids (zeroes inside cells):

$$E + V = n_1^{out} + n_0^{in} \tag{4}$$

The values of these objective functions for the solutions in Tables 2 and 3 are shown below.

$$\eta = \frac{1}{2} \cdot \frac{16}{19} + \frac{1}{2} \cdot \frac{12}{16} \approx 79.60\% \qquad \eta = \frac{1}{2} \cdot \frac{15}{19} + \frac{1}{2} \cdot \frac{11}{16} \approx 73.85\%$$

$$\tau = \frac{20 - 4}{20 + 3} \approx 69.57\% \qquad \tau = \frac{20 - 5}{20 + 4} \approx 62.50\%$$

$$GCI = \frac{20 - 4}{20} \approx 80.00\% \qquad GCI = \frac{20 - 5}{20} \approx 75.00\%$$

$$E + V = 4 + 3 = 7 \qquad E + V = 5 + 4 = 9$$

In this paper, we use the grouping efficiency measure and compare our computational results with the results of [17, 47].

The mathematical programming model of the CFP with the grouping efficiency objective function can be described using boolean variables x_{ik} and y_{jk}. Variable x_{ik} takes value 1 if machine i belongs to cell k and takes value 0 otherwise. Similarly variable y_{jk} takes value 1 if part j belongs to cell k and takes value 0 otherwise. Machines index i takes values from 1 to m and parts index j - from 1 to p. Cells index k takes values from 1 to $c = \min(m, p)$ because every cell should contain at least one machine and one part, and so the number of cells cannot be greater than m and p. Note, that if a CFP solution has n cells then for k from $n + 1$ to c all variables x_{ik}, y_{jk} will be zero in this model. So, we can consider that the CFP solution always has c cells, but some of them can be empty. The mathematical programming formulation is as follows:

$$\max \left(\frac{n_1^{in}}{2n^{in}} + \frac{n_0^{out}}{2n^{out}} \right) \tag{5}$$

where

$$n^{in} = \sum_{k=1}^{c} \sum_{i=1}^{m} \sum_{j=1}^{p} x_{ik} y_{jk}, \quad n^{out} = mp - n^{in}$$

$$n_1^{in} = \sum_{k=1}^{c} \sum_{i=1}^{m} \sum_{j=1}^{p} a_{ij} x_{ik} y_{jk}, \quad n_0^{out} = n_0 - (n^{in} - n_1^{in})$$

subject to

$$\sum_{k=1}^{c} x_{ik} = 1 \quad \forall i \in 1, ..., m \tag{6}$$

$$\sum_{k=1}^{c} y_{jk} = 1 \quad \forall j \in 1, ..., p \tag{7}$$

$$\sum_{i=1}^{m} \sum_{j=1}^{p} x_{ik} y_{jk} \geq \sum_{i=1}^{m} x_{ik} \quad \forall k \in 1, ..., c \tag{8}$$

$$\sum_{i=1}^{m} \sum_{j=1}^{p} x_{ik} y_{jk} \geq \sum_{j=1}^{p} y_{jk} \quad \forall k \in 1, ..., c \tag{9}$$

$$x_{ik} \in \{0, 1\} \quad \forall i \in 1, ..., m \tag{10}$$

$$y_{jk} \in \{0, 1\} \quad \forall j \in 1, ..., p \tag{11}$$

The objective function (5) is the grouping efficiency in this model. Constraints (6) and (7) impose that every machine and every part belongs to some cell. Constraints (8) and (9) guarantee that every nonempty cell contains at least one machine and one part. Note that if singleton cells are not allowed then the right sides of inequalities (8) and (9) should have a coefficient of 2. All these constraints can be linearized in a standard way, but the objective function will still be fractional. That is why the exact solution of this problem presents considerable difficulties.

A cells configuration in the mathematical model is described by the number of machines m_k and parts p_k in every cell k.

$$m_k = \sum_{i=1}^{m} x_{ik}, \quad p_k = \sum_{j=1}^{p} y_{jk}$$

It is easy to see that when a cells configuration is fixed all the optimization criteria (1)–(4) become equivalent (Proposition 1).

Proposition 1 *If a cells configuration is fixed then objective functions (1)–(4): η, τ, GCI, $E+V$ become equivalent and reach the optimal value on the same solutions.*

Proof When a cells configuration is fixed the following values are constant: m_k, p_k. The values of n_1 and n_0 are always constant. The values of n^{in} and n^{out} are constant since $n^{in} = \sum_{k=1}^{c} m_k p_k$ and $n^{out} = mp - n^{in}$. So, if we maximize the number of ones inside the cells n_1^{in} then simultaneously $n_0^{in} = n^{in} - n_1^{in}$ is minimized, $n_0^{out} = n_0 - n_0^{in}$ is maximized, and $n_1^{out} = n_1 - n_1^{in}$ is minimized. This means that the grouping efficiency $\eta = q\frac{n_1^{in}}{n^{in}} + (1-q)\frac{n_0^{out}}{n^{out}}$ is maximized, the grouping efficacy $\tau = \frac{n_1^{in}}{n_1 + n_0^{in}}$ is maximized, the grouping capability index $GCI = 1 - \frac{n_1^{out}}{n_1}$ is maximized, and the number of exceptions plus voids $E + V = n_1^{out} + n_0^{in}$ is minimized simultaneously on the same optimal solution. □

3 Algorithm Description

The main function of our heuristic is presented by algorithm 1.

Algorithm 1 Main function

```
function SOLVE( )
    FINDOPTIMALCELLRANGE(MinCells, MaxCells)
    ConfigsNumber = 2000
    AllConfigs = GENERATECONFIGS(MinCells, MaxCells, ConfigsNumber)
    return CMHEURISTIC(AllConfigs)
end function
```

First we call FINDOPTIMALCELLRANGE($MinCells$, $MaxCells$) function that returns a potentially optimal range of cells - from $MinCells$ to $MaxCells$. Then these values and $ConfigsNumber$ (the number of cell configurations to be generated) are passed to GENERATECONFIGS($MinCells$, $MaxCells$, $ConfigsNumber$) function which generates random cell configurations. The generated configurations $AllConfigs$ are passed to CMHEURISTIC($AllConfigs$) function which finds a high-quality solution for every cell configuration and then chooses the solution with the greatest efficiency value.

Algorithm 2 Procedure for finding the optimal cell range

```
function FINDOPTIMALCELLRANGE( MinCells, MaxCells)
    if (m > p) then
        minDimension = p
    else
        minDimension = m
    end if
    ConfigsNumber = 500
    Configs = GENERATECONFIGS(2, minDimension, ConfigsNumber)
    Solution = CMHEURISTIC(Configs)
    BestCells = GETCELLSNUMBER(Solution)
    MinCells = BestCells - [minDimension * 0,1 ]        ▷ [ ] - integer part
    MaxCells = BestCells + [minDimension * 0,1 ]
end function
```

In function FINDOPTIMALCELLRANGE($MinCells, MaxCells$) (Algorithm 2) we look over all the possible number of cells from 2 to maximal possible number of cells which is equal to $\min(m, p)$. For every number of cells in this interval, we generate a fixed number of configurations (we use 500in this paper) calling GENERATECONFIGS(2, $minDimension, ConfigsNumber$) and then use our CMHEURISTIC($Configs$) to obtain a potentially optimal number of cells. But we consider not only one number of cells but together with its 10%-neighborhood $[MinCells, MaxCells]$.

Algorithm 3 Configurations generation

```
function GENERATECONFIGS(MinCells, MaxCells, ConfigsNumber)
    Configs = ∅
    for cells = MinCell, MaxCells do
        Generated = GENERATECONFIGS(cells, ConfigsNumber)
        Configs = Configs ∪ Generated
        return Configs
    end for
end function
```

Function GENERATECONFIGS($MinCells, MaxCells, ConfigsNumber$) (Algorithm 3) returns a set of randomly generated cell configurations with a number of cells ranging from $MinCells$ to $MaxCells$. We call GENERATECONFIGSUNIFORM($cells, ConfigsNumber$) function which randomly selects with uniform distribution $ConfigsNumber$ configurations from all possible cell configurations with the specified number of cells. Note that mathematically a cell configuration with k cells can be represented as an integer partition of m and p values into sums of k summands. We form a set of configurations for every number of cells and then join them.

Algorithm 4 CMHeuristic

```
function CMHEURISTIC(Configs)
    Best = 0
    for all config ∈ Configs do
        Solution = IMPROVESOLUTION(config)
        if Solution > Best then
            Best = Solution
        end if
    end for
    return Best
end function
```

Function CMHEURISTIC(Configs) (Algorithm 4) gets a set of cell configurations and for each configuration runs an improvement algorithm to obtain a good solution. A solution includes a permuted machine-part matrix, a cell configuration, and the corresponding grouping efficiency value. The function chooses the best solution and returns it.

Improvement procedure IMPROVESOLUTION($config$, $\eta_{current}$) (Algorithm 5) works as follows. We consider all the machines and the parts in order to know if there is a machine or a part that we can move to another cell and improve the current efficiency $\eta_{current}$. First we consider moving of every part on all other cells and compute how the efficiency value changes. Here $\eta_{part,cell}$ is the efficiency of the current solution where the part with index $part$ is moved to the cell with index $cell$. This operation is performed for all the parts and the part with the maximum increase in efficiency Δ_{parts} is chosen. Then we repeat the same operations for all the machines. Finally, we compare the best part movement and the best machine movement and choose the one with the highest efficiency. This procedure is performed until any improvement is possible and after that we get the final solution.

The main idea of IMPROVESOLUTION($config$, $\eta_{current}$) is illustrated on [39] instance 8 × 12 (Table 4). To compute the grouping efficiency for this solution, we need to know the number of ones inside cells n_1^{in}, the total number of elements inside

Table 4 [39] instance 8 × 12

	1	2	3	4	5	6	7	8	9	10	11	12
1	1	1	1	1	0	0	0	0	0	0	0	0
2	1	0	1	1	1	1	1	0	0	1	0	0
3	0	0	1	1	1	1	1	1	1	0	0	0
4	0	0	0	0	0	1	1	1	1	1	0	0
5	0	0	0	0	0	0	1	1	1	1	0	0
6	0	0	0	0	0	0	1	1	1	0	1	0
7	0	0	0	0	0	0	0	0	0	0	1	1
8	0	0	0	0	0	0	0	0	0	0	1	1

Algorithm 5 Solution improvement procedure

```
function IMPROVESOLUTION(config, η_current)
    η_current = GROUPINGEFFICIENCY(config)
    repeat
        PartFrom = 0
        PartTo = 0
        for part = 1, partsNumber do
            for cell = 1, cellsNumber do
                if (η_part,cell > η_current) then
                    Δ_parts = (η_part,cell − η_current)
                    PartFrom = GetPartCell(part)
                    PartTo = cell
                end if
            end for
        end for
        MachineFrom = 0
        MachineTo = 0
        for machine = 1, machinesNumber do
            for cell = 1, cellsNumber do
                if (η_machine,cell > η_current) then
                    Δ_machines = (η_machine,cell − η_current)
                    MachineFrom = GETMACHINECELL(machine)
                    MachineTo = cell
                end if
            end for
        end for
        if Δ_parts > Δ_machines then
            MOVEPART(PartFrom, PartTo)
        else
            MOVEMACHINE(MachineFrom, MachineTo)
        end if
    until Δ > 0
end function
```

cells n^{in}, the number of zeros outside cells n_0^{out}, and the number of elements outside cells n^{out}. The grouping efficiency is then calculated by the following formula:

$$\eta = q \cdot \frac{n_1^{in}}{n^{in}} + (1-q) \cdot \frac{n_0^{out}}{n^{out}} = \frac{1}{2} \cdot \frac{20}{33} + \frac{1}{2} \cdot \frac{48}{63} \approx 68.4\%$$

Looking at this solution (Table 4) we can conclude that it is possible, for example, to move part 4 from the second cell to the first one. And this way, the number of zeros inside cells decreases by 3 and the number of ones outside cells also decreases by 4. So, it is profitable to attach column 4 to the first cell as it is shown on Table 5. For the modified cells configuration we have:

$$\eta = \frac{1}{2} \cdot \frac{23}{33} + \frac{1}{2} \cdot \frac{51}{63} \approx 75.32\%$$

Table 5 Moving part 4 from cell 2 to cell 1

	1	2	3	4	5	6	7	8	9	10	11	12
1	1	1	1	1	0	0	0	0	0	0	0	0
2	1	0	1	1	1	1	1	0	0	1	0	0
3	0	0	1	1	1	1	1	1	1	0	0	0
4	0	0	0	0	0	1	1	1	1	1	0	0
5	0	0	0	0	0	0	1	1	1	1	0	0
6	0	0	0	0	0	0	1	1	1	0	1	0
7	0	0	0	0	0	0	0	0	0	0	1	1
8	0	0	0	0	0	0	0	0	0	0	1	1

Table 6 Maximal efficiency increase for each row

	1	2	3	4	5	6	7	8	9	10	11	12	
1	1	1	1	1	0	0	0	0	0	0	0	0	−6.94%
2	1	0	1	1	1	1	1	0	0	1	0	0	+1.32%
3	0	0	1	1	1	1	1	1	1	0	0	0	+7.99%
4	0	0	0	0	0	1	1	1	1	1	0	0	−0.07%
5	0	0	0	0	0	0	1	1	1	1	0	0	+0.77%
6	0	0	0	0	0	0	1	1	1	0	1	0	+0.77%
7	0	0	0	0	0	0	0	0	0	0	1	1	−4.62%
8	0	0	0	0	0	0	0	0	0	0	1	1	−4.62%

As a result the efficiency is increased almost for 7%. Computational results show that using such modifications could considerably improve the solution. The idea is to compute an increase in efficiency for each column and row when it is moved to another cell and then perform the modification corresponding to the maximal increase. For example, Table 6 shows the maximal possible increase in efficiency for every row when it is moved to another cell.

4 Computational Results

In all the experiments for determining a potentially optimal range of cells we use 500 random cell configurations for each cells number, and for obtaining the final solution we use 2000 random configurations. An Intel Core i7 machine with 2.20 GHz CPU and 8.00 Gb of memory is used in our experiments. We run our heuristic on 24 CFP benchmark instances taken from the literature. The sizes of the considered problems vary from 10×20 to 50×150. The computational results are presented in Table 7. For every instance we make 50 algorithm runs and report minimum, average, and maximum value of the grouping efficiency obtained by the suggested heuristic

Table 7 Computational results

#	Source	mxp	Efficiency value, %					Time, sec	Cells
			Bhatnagar and Saddikuti	Goldengorin et al.	Our				
					Min	Avg	Max		
1	Sandbothe [38]	10×20	95.40	95.93[a]	95.66	95.66	**95.66**	0.36	7
2	Ahi et al. [1]	20×20	92.62	93.85	95.99	95.99	**95.99**	0.62	9
3	Mosier and Taube [33]	20×20	85.63	88.71	90.11	90.16	**90.22**	0.88	9
4	Boe and Cheng [4]	20×35	88.31	88.05	93.34	93.47	**93.55**	1.62	10
5	Carrie [8]	20×35	90.76	95.64	95.43	95.78	**95.79**	1.54	10
6	Ahi et al. [1]	20×51	87.86	94.11	95.36	95.4	**95.45**	3.1	12
7	Chandrasekharan et al. [11]	20×40	98.82	100.00	100	100	100	1.8	7
8	Chandrasekharan et al. [11]	20×40	95.33	97.48	97.7	97.75	**97.76**	2.42	12
9	Chandrasekharan et al. [11]	20×40	93.78	96.36	96.84	96.88	**96.89**	2.56	12
10	Chandrasekharan et al. [11]	20×40	87.92	94.32	96.11	96.16	**96.21**	3.3	15
11	Chandrasekharan et al. [11]	20×40	84.95	94.21	95.94	96.03	**96.1**	2.84	15
12	Chandrasekharan et al. [11]	20×40	85.06	92.32	95.85	95.9	**95.95**	2.76	15
13	Nair and Narendran [34]	20×40	96.44	97.39	97.78	97.78	**97.78**	2.12	10
14	Nair and Narendran [34]	20×40	92.35	95.74	97.4	97.4	**97.4**	2.2	14
15	Nair and Narendran [34]	20×40	93.25	95.70	95.81	96.03	**96.17**	2.48	12
16	Nair and Narendran [34]	20×40	91.11	96.40	96.98	96.98	**96.98**	2.78	14
17	Ahi et al. [1]	25×40	91.09	95.52	96.48	96.48	**96.48**	2.58	14
18	Yang and Yang [47]	28×35	93.43	93.82	94.81	94.85	**94.86**	2.46	10
19	Kumar and Vannelli [24]	30×41	90.66	97.22	97.38	97.53	**97.62**	3.54	18
20	Stanfel [42]	30×50	88.17	96.48	96.77	96.83	**96.9**	5.02	18

(continued)

Table 7 (continued)

#	Source	mxp	Efficiency value, %					Time, sec	Cells
			Bhatnagar and Saddikuti	Goldengorin et al.	Our				
					Min	Avg	Max		
21	King and Nakornchai [21]	30 × 90	83.18	94.62	95.37	95.84	**96.27**	13.1	25
22	Chandrasekharan and Rajagopalan [10]	40 × 100	94.75	95.91	98.06	98.1	**98.13**	16.88	17
23	Yang and Yang [47]	46 × 105	90.98	95.20	96.1	96.18	**96.29**	23.9	18
24	Liang and Zolfaghari [27]	50 × 150	93.05	92.92	96.08	96.17	**96.27**	51.66	24

[a]This solution has a mistake

over these 50 runs. We compare our results with the best known values taken from [3, 17]. We have found better solutions unknown before for 23 instances of the 24 considered. For CFP instance 6, we have found the same optimal solution with 100% of grouping efficiency as in [17]. For CFP instance 1 the solution of [17] has some mistake. For this instance having a small size of 10×20 it can be proved that our solution is the global optimum applying an exact approach [14] for the grouping efficiency objective and all the possible number of cells from 1 to 10.

5 Concluding Remarks

In this paper, we present a new heuristic algorithm for solving the CFP. The high quality of the solutions is achieved due to the enumeration of different numbers of cells and different cell configurations and applying our improvement procedure. Since the suggested heuristic works fast (the solution for one cell configuration is achieved in several milliseconds for any instance from 10×20 to 50×150), we apply it for thousands of different configurations. Thus a big variety of good solutions is covered by the algorithm and the best of them has high grouping efficiency.

Acknowledgements This work was conducted at the Laboratory of Algorithms and Technologies for Network Analysis of National Research University Higher School of Economics and partly supported by RSF 14-41-00039 grant.

References

1. Ahi, A., Mir, B., Aryanezhad, B., Ashtiani, A.M.: A novel approach to determine cell formation, intracellular machine layout and cell layout in the CMS problem based on TOPSIS method. Comput. Oper. Res. **36**(5), 1478–1496 (2009)
2. Askin, R.G., Chiu, K.S.: A graph partitioning procedure for machine assignment and cell formation in group technology. Int. J. Prod. Res. **28**(8), 1555–1572 (1990)
3. Bhatnagar, R., Saddikuti, V.: Models for cellular manufacturing systems design: matching processing requirements and operator capabilities. J. Opl. Res. Soc. **61**, 827–839 (2010)
4. Boe, W., Cheng, C.H.: A close neighbor algorithm for designing cellular manufacturing systems. Int. J. Prod. Res. **29**(10), 2097–2116 (1991)
5. Burbidge, J.L.: The new approach to production. Prod. Eng. (1961). (Dec. 3–19)
6. Burbidge, J.L.: The Introduction of Group Technology. Wiley, New York (1975)
7. Bychkov, I., Batsyn, M., Sukhov, P.: Ch. 4., Heuristic algorithm for the cell formation problem. In: Goldengorin, B.I., Kalyagin, V.A., Pardalos, P.M. (eds.) Models, Algorithms, and Technologies for Network Analysis, vol. 59, pp. 43–69. Springer, NY (2013)
8. Carrie, S.: Numerical taxonomy applied to group technology and plant layout. Int. J. Prod. Res. **11**, 399–416 (1973)
9. Chandrasekaran, M.P., Rajagopalan, R.: MODROC: an extension of rank order clustering of group technology. Int. J. Prod. Res. **24**(5), 1221–1233 (1986)
10. Chandrasekharan, M.P., Rajagopalan, R.: ZODIAC–an algorithm for concurrent formation of part families and machine cells. Int. J. Prod. Res. **25**(6), 835–850 (1987)

11. Chandrasekharan, M.P., Rajagopalan, R.: Groupability: analysis of the properties of binary data matrices for group technology. Int. J. Prod. Res. **27**(6), 1035–1052 (1989)
12. Chan, H.M., Milner, D.A.: Direct clustering algorithm for group formation in cellular manufacture. J. Manuf. Syst. **1**(1), 64–76 (1982)
13. Cheng, C.H., Gupta, Y.P., Lee, W.H., Wong, K.F.: A TSP-based heuristic for forming machine groups and part families. Int. J. Prod. Res. **36**(5), 1325–1337 (1998)
14. Elbenani, B., Ferland, J.A.: Manufacturing Cell Formation by State-Space Search, pp. 14. Montreal, Quebec. CIRRELT-2012-07 (2012)
15. Flanders, R.E.: Design manufacture and production control of a standard machine. Trans. ASME **46**, 691–738 (1925)
16. Ghosh, S., Mahanti, A., Nagi, R., Nau, D.S.: Cell formation problem solved exactly with the dinkelbach algorithm. Ann. Oper. Res. **65**(1), 35–54 (1996)
17. Goldengorin, B., Krushinsky, D., Slomp, J.: Flexible PMP approach for large size cell formation. Oper. Res. **60**(5), 1157–1166 (2012)
18. Goncalves, J.F., Resende, M.G.C.: An evolutionary algorithm for manufacturing cell formation. Comput. Ind. Eng. **47**, 247–273 (2004)
19. Hsu, C.P.: Similarity coefficient approaches to machine-component cell formation in cellular manufacturing: a comparative study. Ph.D. Thesis, Department of Industrial and Manufacturing Engineering, University of Wisconsin Milwaukee (1990)
20. King, J.R.: Machine-component grouping in production flow analysis: an approach using a rank order clustering algorithm. Int. J. Prod. Res. **18**(2), 213–232 (1980)
21. King, J.R., Nakornchai, V.: Machine-component group formation in group technology: review and extension. Int. J. Prod. Res. **20**(2), 117–133 (1982)
22. Krushinsky, D., Goldengorin, B.: An exact model for cell formation in group technology. CMS **9**, 323–338 (2012). doi:10.1007/s10287-012-0146-2
23. Kumar, K.R., Chandrasekharan, M.P.: Grouping efficacy: a quantitative criterion for goodness of block diagonal forms of binary matrices in group technology. Int. J. Prod. Res. **28**(2), 233–243 (1990)
24. Kumar, K.R., Vannelli, A.: Strategic subcontracting for efficient disaggregated manufacturing. Int. J. Prod. Res. **25**(12), 1715–1728 (1987)
25. Kusiak, A.: The generalized group technology concept. Int. J. Prod. Res. **25**(4), 561–569 (1987)
26. Lei, D., Wu, Z.: Tabu search for multiple-criteria manufacturing cell design. Int. J. Adv. Manuf. Technol. **28**, 950–956 (2006)
27. Liang, M., Zolfaghari, S.: Machine cell formation considering processing times and machine capacities: an ortho-synapse Hopfield neural netowrk approach. J. Intell. Manuf. **10**, 437–447 (1999)
28. McAuley, J.: Machine grouping for efficient production. Prod. Eng. Res. Dev. **51**(2), 53–57 (1972)
29. McCormick, W.T., Schweitzer, P.J., White, T.W.: Problem decomposition and data reorganization by a clustering technique. Oper. Res. **20**(5), 993–1009 (1972)
30. Mitrofanov, S.P.: Nauchnye osnovy gruppovoy tekhnologii. Lenizdat, Leningrad, Russia, 435 pp (1959) (in Russian)
31. Mitrofanov, S.P.: The Scientific Principles of Group Technology. Boston Spa. National Lending Library Translation, Yorkshire (1966) (Translation of Mitrofanov (1959))
32. Mosier, C.T.: An experiment investigating the application of clustering procedures and similarity coefficients to the GT machine cell formation problem. Int. J. Prod. Res. **27**(10), 1811–1835 (1989)
33. Mosier, C.T., Taube, L.: Weighted similarity measure heuristics for the group technology machine clustering problem. OMEGA **13**(6), 577–583 (1985)
34. Nair, G.J.K., Narendran, T.T.: Grouping index: a new quantitative criterion for goodness of block-diagonal forms in group technology. Int. J. Prod. Res. **34**(10), 2767–2782 (1996)
35. Ng, S.: Worst-case analysis of an algorithm for cellular manufacturing. Eur. J. Oper. Res. **69**(3), 384–398 (1993)

36. Ng, S.: On the characterization and measure of machine cells in group technology. Oper. Res. **44**(5), 735–744 (1996)
37. Rajagopalan, R., Batra, J.L.: Design of cellular production systems: a graph-theoretic approach. Int. J. Prod. Res. **13**(6), 567–579 (1975)
38. Sandbothe, R.A.: Two observations on the grouping efficacy measure for goodness of block diagonal forms. Int. J. Prod. Res. **36**, 3217–3222 (1998)
39. Seifoddini, H., Wolfe, P.M.: Application of the similarity coefficient method in group technology. IIE Trans. **18**(3), 271–277 (1986)
40. Shtub, A.: Modelling group technology cell formation as a generalized assignment problem. Int. J. Prod. Res. **27**(5), 775–782 (1989)
41. Srinivasan, G., Narendran, T.T.: GRAFICS-A nonhierarchical clustering-algorithm for group technology. Int. J. Prod. Res. **29**(3), 463–478 (1991)
42. Stanfel, L.: Machine clustering for economic production. Eng. Costs Prod. Econ. **9**, 73–81 (1985)
43. Waghodekar, P.H., Sahu, S.: Machine-component cell formation in group technology MACE. Int. J. Prod. Res. **22**, 937–948 (1984)
44. Won, Y., Lee, K.C.: Modified p-median approach for efficient GT cell formation. Comput. Ind. Eng. **46**, 495–510 (2004)
45. Wu, X., Chao-Hsien, C., Wang, Y., Yan, W.: A genetic algorithm for cellular manufacturing design and layout. Eur. J. Oper. Res. **181**, 156–167 (2007)
46. Xambre, A.R., Vilarinho, P.M.: A simulated annealing approach for manufacturing cell formation with multiple identical machines. Eur. J. Oper. Res. **151**, 434–446 (2003)
47. Yang, M.-S., Yang, J.-H.: Machine-part cell formation in group technology using a modified ART1 method. Eur. J. Oper. Res. **188**(1), 140–152 (2008)

Efficient Methods of Multicriterial Optimization Based on the Intensive Use of Search Information

Victor Gergel and Evgeny Kozinov

Abstract In this paper, an efficient approach for solving complex multicriterial optimization problems is proposed. For the problems being solved, the optimality criteria may be multiextremal ones, and calculating the criteria values may require a large amount of computations. The proposed approach is based on reducing multicriterial problems to nonlinear programming problems via the minimax convolution of the partial criteria, reducing dimensionality by using Peano evolvents, and applying efficient information-statistical global optimization methods. The new contribution is that all the search information obtained in the course of optimization is used to find each current Pareto-optimal solution. The results of the computational experiments show that the proposed approach essentially reduces the computational costs of solving multicriterial optimization problems (by tens and hundreds of times).

1 Introduction

Multicriterial optimization (MCO) problems are the subject of intense research and are widely used in applications. The practical demand stimulates extensive research in the field of MCO problems—see, for example, monographs [3, 5, 20, 22, 24] and reviews of the scientific and practical results [6, 8, 17, 21, 23, 28, 34]. As a result, a great number of efficient methods for solving MCO problems have been proposed, and the solutions to many practical problems have been reported.

Possible contradictions between the partial efficiency criteria are meaningful issues with multicriterial optimization problems. As a result, finding the optimal (best) values for all partial criteria simultaneously is impossible. As a rule, improving the efficiency with respect to some partial criteria results in the reduced quality of chosen solutions with respect to other criteria. In these situations, solving a MCO

V. Gergel (✉) · E. Kozinov
Nizhni Novgorod State University, 23, Gagarin av., Nizhny Novgorod, Russia
e-mail: gergel@unn.ru

E. Kozinov
e-mail: evgeny.kozinov@itmm.unn.ru

V.A. Kalyagin et al. (eds.), *Models, Algorithms, and Technologies for Network Analysis*, Springer Proceedings in Mathematics & Statistics 197,
DOI 10.1007/978-3-319-56829-4_3

problem consists in finding some compromise solutions, for which the obtained values are coordinated with respect to partial criteria. It is worth noting that the concept of an expedient compromise may change during the course of computations, which may require finding several different compromise solutions.

This work is dedicated to solving MCO problems which are used to describe decision-making problems for designing complex engineered devices and systems. In such applications, the partial criteria may take a complex multiextremal form, and the computations of the criteria values are computationally expensive procedures as a rule. In these conditions, finding even one compromise solution requires a significant number of computations, whereas determining several Pareto-efficient solutions (or the entire set of them) becomes a challenging problem. In order to overcome this problem, maximizing the use of search information obtained during the course of computations is proposed. Within this approach, finding every next compromise solution requires fewer and fewer computations down to executing just a few iterations to find the next efficient solution.

This article is organized as follows. In Sect. 2, the multicriterial optimization problem statement is given. In Sect. 3, the basics of the approach are presented, namely reducing multicriterial problems to nonlinear programming problems using the minimax convolution of partial criteria, and reducing dimensionality using Peano evolvents. In Sect. 4, the multidimensional generalized global search algorithm is described for solving the reduced scalar nonlinear programming problems is described, and issues are discussed regarding the reuse of search information obtained during the course of computations. Section 5 includes results from computational experiments. In the Conclusion, the obtained results are discussed, and the main areas for further investigation are presented.

2 Problem Statement

A multicriterial optimization (MCO) problem can be defined as follows:

$$f(y) = (f_1(y), f_2(y), \dots, f_s(y)) \rightarrow \min, \ y \in D, \tag{1}$$

where:

- $y = (y_1, y_2, \dots, y_N)$ is the vector of varied parameters,
- N is the dimensionality of the multicriterial optimization problem being solved,
- D is the search domain being an N-dimensional hyperparallelepiped

$$D = \{y \in R^N : a_i \leq y_i \leq b_i, 1 \leq i \leq N\}$$

at a given boundary of vectors a and b. Without any loss of generality, the partial criteria values in problem (1) are supposed to be non-negative, and reducing these corresponds to an increase in the efficiency of the considered solutions $y \in D$.

The partial criteria in MCO problem (1) are usually contradictory, and there is no solution $y \in D$ that provides optimal values for all criteria simultaneously. In these cases, such solutions $y^* \in D$ are considered to be the solutions of the MCO problem, for which the values of any particular criteria cannot be improved without reducing the efficiency with respect to other criteria. Such un-improvable solutions are called *efficient* or *Pareto-optimal*. Any efficient solution can be considered a *partial solution*, whereas the set of all un-improvable solutions is the *complete solution* of the MCO problem.

As mentioned above, in this work, problem (1) will be applied to the most complex decision-making problems, for which the partial criteria $f_i(y)$, $1 \le i \le s$, can be multiextremal, and obtaining criteria values at the points of the search domain $y \in D$ may require a considerable number of computations. Let us also assume that the partial criteria $f_i(y)$, $1 \le i \le s$, satisfy the Lipschitz condition

$$|f_i(y') - f_i(y'')| \le L_i \|y' - y''\|,\ y', y'' \in D,\ 1 \le i \le s. \tag{2}$$

where L_i is the Lipschitz constant for the criterion $f_i(y)$, $1 \le i \le s$.

It is important to note that the feasibility of the Lipschitz condition fits practical applications well—at small variations in the parameter $y \in D$, the corresponding changes of the partial criteria values are limited as a rule.

3 The Basics of the Approach

3.1 Methods of Solving the Multicriterial Optimization Problems

Multicriterial optimization is a field of intensive scientific investigations. Among the approaches developed for solving MCO problems, one can select a *lexicographical optimization* method where the criteria are arranged in a certain way according to their importance, and the optimization of partial criteria is performed step by step as the level of importance decreases—see, for example, [3]. *Interactive methods* [2, 21] represent another approach where the researcher (decision maker, DM) is involved in the process of choosing solutions. Another extensively developed area is the development of *evolutionary algorithms* based on the imitation of certain natural phenomena for applying them to solving MCO problems [2, 4, 31, 33].

The *scalarization approach*, in which some convolution methods of a set of partial criteria $f_i(y)$, $1 \leq i \leq s$, are applied to an integrated scalar functional $F(\lambda, y)$, is an extensively developed area for solving MCO problems—see, for example [5, 6]. Such an approach reduces the solution of problem (1) to solving a nonlinear programming problem

$$\min F(\lambda, y), y \in D \tag{3}$$

where $\lambda = (\lambda_1, \lambda_2, \ldots, \lambda_s)$ is a vector of coefficients used to construct integrated scalar criterion. As part of this approach, a wide set of scalarization methods have been proposed for partial criteria. From among them, one can select various kinds of convolution, including the additive, minimax, and multiplicative schemes.

Various methods for defining the preferred solutions, which should be obtained as a result of solving MCO problems, can also lead to scalar criterion. Among such approaches are methods for seeking solutions which are closest to the ideal or to compromise solutions or to existing prototypes, etc. More detailed consideration of this approach in given, for example, in [5, 21, 24].

The coefficients λ from (3), used in the approaches listed above, are often the requirements for an expedient compromise combination of the partial criteria. Thus, for example, the scalar criterion in the minimax convolution is defined as

$$F(\lambda, y) = \max_{1 \leq i \leq s} \lambda_i f_i(y) \tag{4}$$

where the coefficients λ_i, $1 \leq i \leq s$ should be non-negative, and their sum should be balanced to the unit value:

$$\sum_{i=1}^{s} \lambda_i = 1, \lambda_i \geq 0, 1 \leq i \leq s.$$

The necessity and sufficiency of this approach to solving MCO problems is one of the main properties of the minimax convolution scheme: the results of minimizing $F(\lambda, y)$ lead to obtaining efficient solutions[1] to MCO problems, and, vice versa, any efficient solution of the MCO problem can be obtained as a result of minimizing $F(\lambda, y)$ at the corresponding values of the convolution coefficients λ_i, $1 \leq i \leq s$—see, for example, [21].

The coefficients λ_i, $1 \leq i \leq s$ in (4) can be considered indicators of the importance of the partial criteria—the greater the value of coefficient λ_i for some partial criterion, the greater the contribution of this partial criterion to the integrated scalar criterion $F(\lambda, y)$. Therefore, a method for solving MCO problems where the compromise solution sought is determined during the course of several stages performed sequentially. At every stage, the researcher (decision maker) specifies the desired

[1]More precisely, the minimization of $F(\lambda, y)$ can lead to obtaining weakly efficient solutions (the set of weakly efficient solutions includes the Pareto domain). The situation can be corrected by adding an additional correcting element into (4)—see, for example, [21].

values of the importance coefficients λ_i, $1 \le i \le s$ then problem (4) formulated in this way is solved. After that, the researcher analyzes the efficient solutions obtained and corrects the specified coefficients λ_i, $1 \le i \le s$ if necessary. Such a multistep method corresponds to the practice of decision-making for compromise solutions in complex optimization problems. The possibility of determining several efficient solutions (or an entire set) at a reasonable computation cost becomes a key issue in solving complex multicriterial optimization problems.

3.2 Dimensionality Reduction

In the general case, finding numerical estimates for globally optimized solutions implies generating coverage of the search domain D—see, for example, [7, 9, 18, 19, 25, 32, 35, 36]. As a result, the computational costs of solving global optimization problems are very high, even with a relatively low number of varied parameters (the dimensionality of the problem). A considerable decrease in the number of computations can be achieved if the computing grids obtained when covering the search domain are non-uniform, when the optimization points are only denser in close proximity to the globally optimized solutions. The constricting of such economic non-uniform coverages considerably complicates the computational schemes of global optimization methods. One possible way to reduce this complexity involves using various dimensionality reduction methods [25, 27, 29, 30].

Within the framework of the information-statistical theory of global optimization, Peano *curves* or *evolvents* $y(x)$ mapping the interval [0, 1] onto the N-dimensional hypercube D unambiguously are used for dimensionality reduction—see, for example [27, 29, 30]. As a result of this reduction, the initial multidimensional global optimization problem (4) is reduced to a one-dimensional problem:

$$F(\lambda, y(x^*)) = \min\{F(\lambda, y(x)) : x \in [0, 1]\}. \tag{5}$$

The dimensionality reduction scheme considered associates a multidimensional problem with a Lipschitzian minimized function with a series of one-dimensional problems, for which the corresponding objective functions satisfy the uniform Hölder condition (see [29, 30]) i.e.

$$|F(\lambda, y(x')) - F(\lambda, y(x''))| \le H|x' - x''|^{1/N}, x', x'' \in [0, 1], \tag{6}$$

where the Hölder constant H is defined by the relationship $H = 4L\sqrt{N}$, where L is the Lipschitz constant of the function $F(\lambda, y)$, and N is the dimensionality of the optimization problem (4).

It can be noted that a nested optimization scheme can also be applied for dimensionality reduction—see [1, 11, 12, 30].

4 An Efficient Method for Solving the Multicriterial Optimization Problems Based on Reusing Search Information

The basics of the approach presented in Sect. 3 allow the solution of the MCO problem (1) to be reduced to solving a series of reduced multiextremal problems (5). And, therefore, the problem of developing methods for solving the MCO problem is resolved by the potentially broad application of global search algorithms.

4.1 Method for Solving Global Optimization Problems

It should be pointed that multiextremal optimization is an area of extensive research—the general state of the art is presented, for example, in [9, 18, 19, 25, 30, 32, 35], etc. The main results from applying dimensionality reduction using Peano evolvents have been obtained through the information-statistical theory of global search developed in [29, 30]. This theory has served as the basis for developing a large number of efficient methods for multiextremal optimization—see, for example [1, 11–15, 26], etc.

Within the framework of this approach, the Generalized Multidimensional Algorithm of Global Search (GMAGS) [13, 29, 30] forms the basis for the optimization methods being developed. The general computational scheme of this algorithm can be presented as follows.

Let us introduce a simpler notation for reduced one-dimensional problems (5) as

$$\phi(x) = F(\lambda, y(x)) : x \in [0, 1]. \tag{7}$$

The initial iteration of the algorithm is performed at an arbitrary point $x^1 \in (0, 1)$. Then, let us assume $k, k > 1$ global search iterations to be completed. The choice of the optimization point of the next $(k + 1)^{th}$ iteration is determined by the following rules.

Rule 1. Renumber the optimization points by the lower indices in the order of increasing coordinate value

$$0 = x_0 < x_1 < \cdots < x_i < \cdots < x_k < x_{k+1} = 1, \tag{8}$$

the points x_0, x_{k+1} have been introduced additionally for the convenience of further explanation, the values of the minimized function z_0, z_{k+1} at these points are undefined.

Rule 2. Compute the current estimate of the Hölder constant H from (6)

$$m = \begin{cases} rM, & M > 0 \\ 1, & M = 0 \end{cases}, M = \max_{1 \le i \le k} \frac{|z_i - z_{i-1}|}{\rho_i} \tag{9}$$

as the relative difference in the values of the minimized functions $\phi(x)$ from (7) on the set of the points of the executed iterations x_i, $1 \le i \le k$ from (8). Here and hereafter $\varrho_i = (x_i - x_{i-1})^{1/N}, 1 \le i \le k+1$. The constant $r, r > 1$ is the *parameter* for the algorithm.

Rule 3. For each interval (x_{i-1}, x_i), $1 \le i \le k+1$ compute the *characteristic* $R(i)$ where

$$R(i) = \varrho_i + \frac{(z_i - z_{i-1})^2}{m^2 \varrho_i} - 2\frac{(z_i + z_{i-1})}{m}, 1 < i \le k,$$

$$R(i) = 2\varrho_i - 4\frac{z_i}{m}, i = 1, \tag{10}$$

$$R(i) = 2\varrho_i - 4\frac{z_{i-1}}{m}, i = k+1$$

Rule 4. Determine the interval with the maximum characteristic

$$R(t) = \max_{1 \le i \le k+1} R(i) \tag{11}$$

Rule 5. Execute a new trial (computing the value of the minimized function $\phi(x)$) at the point x^{k+1} placed in the interval with the maximum characteristic from (11)

$$x^{k+1} = \frac{x_t + x_{t-1}}{2} - sign(z_t - z_{t-1})\frac{1}{2r}[\frac{|z_t - z_{t-1}|}{m}]^N, 1 < i \le k$$

$$x^{k+1} = \frac{x_t + x_{t-1}}{2}, t = 1, t = k+1. \tag{12}$$

The termination condition, by which the trials are terminated, is defined by the condition

$$\varrho_t < \varepsilon, \tag{13}$$

for the interval t with the maximum characteristic $R(t)$ form (11) and $\varepsilon > 0$ is the given accuracy of the solution. If the termination condition is not fulfilled, the iteration number k is incremented by unity, and a new iteration of the global search is performed.

To clarify the presented algorithm, let us note the following. The concrete form of the characteristics $R(i)$, $1 \le t \le k+1$, calculated in (10) have been developed with the framework of the information-statistical theory of global optimization, and can be interpreted as some measure of the importance of the intervals with respect to containing the global minimum point within them. As it can be seen, the characteristics stimulate the selection of new iteration points within the longest intervals and with the smallest values of the function to be minimized. Then, the scheme for choosing the interval to execute the next trial described in (11)–(12) becomes clear—the point of every next iteration of a global search is chosen in the interval in which the occurrence of the global minimum point is the most probable.

The condition where the presented algorithm converges has been considered, for example, in [30]. Thus, at a proper estimate of the Hölder constant ($m > 2^{2-1/N}H$, m is from (9)) the algorithm converges to all existing global minimum points.

It is worth noting that the results obtained for GMAGS in this work are applicable to the majority of multiextremal optimization methods that can be formulated in accordance with the general characteristic scheme [15]. Moreover, the proposed approach can be extended efficiently to parallel computations—see, for instance [13, 30].

4.2 Reusing Search Information to Efficiently Solve Multicriterial Problems

Solving multicriterial optimization problems may require a large number of computations. The main problem for computation costs consists of the fact that, in general, multiextremal problems (5) must be solved several times. The full usage of all search information obtained during the course of computations could overcome this problem.

The numerical solution of multicriterial optimization problems usually consists of successive computations of the partial criteria values $f^i = f(y^i)$ at the points y^i, $1 \leq i \leq k$, of the search domain D (see, for example, the rules of the GMAGS algorithm). The search information obtained as a result of the computations can be represented as a set (*set of the search information*, SSI):

$$\Omega_k = \{(y^i, f^i = f(y^i))^T : 1 \leq i \leq k\}. \tag{14}$$

It is important to note that SSI contains all of the available information on the optimization problem being solved, and any possible increasing in the efficiency of the global search can be organized based on the information stored in SSI. Given that, the size of the search information when solving complex multidimensional problems may appear to be large enough. However, as a rule, any reduction in the stored data volume results in executing excess global search iterations.

As a result of scalarizing vector criterion (3) or (4), dimensionality reduction (5), and the need for the ordered placement of the trial points (see Rule 1 of GMAGS) SSI is transformed into a *set of optimization data* (SOD)

$$A_k = \{(x_i, z_i, l_i)^T : 0 \leq i \leq k+1\}, \tag{15}$$

where

- $x_i, 0 \leq i \leq k+1$ are the reduced points of the executed global search iterations, in which the criteria values have been computed; the arranged placement of the points according to Rule 1 of GMAGS is reflected by the use of the lower index, i.e.,

$$0 = x_0 < x_1 < \cdots < x_{k+1} = 1;$$

- z_i, $0 \leq i \leq k+1$ are the scalar criterion values for the current optimization problem (5) being solved at the points x_i, $0 \leq i \leq k+1$, i.e.,

$$z_i = \phi(x_i) = F(\lambda, y(x_i)), 1 \le i \le k$$

(the values z_0, z_{k+1}, are undefined and are not used in the computations);

- l_i, $1 \le i \le k$ are the global search iterations indices, for which the points x_i, $1 \le i \le k$ have been computed; these indices are used to store the correspondence between the reduced points of the executed iterations and the multidimensional ones, i.e.,

$$y^j = y(x_i), j = l_i, 1 \le i \le k.$$

In contrast to SSI, the set of optimization data contains the search information reduced to the current scalar reduced problem (5) being solved. In addition, the search information in SOD is represented in a form allowing efficient performing of the global search algorithms. Thus, for example, SOD supports the ordered placement of iteration points necessary for GMAGS—see Rule 1.

The definition of SSI and SOD has a very crucial impact on the essential reduction of the computational cost for solving multicriterial optimization problems. The optimization methods may use SOD to adaptively perform the scheduled search iterations (taking into account the results of previous computations). And, as the main contribution of the proposed approach, the availability of SSI allows the previous computations to be recalculated in SOD for the values of the current optimization problem (5) to be solved without any costly computation of the values for the partial criteria $f_i(y)$, $1 \le i \le s$ from (1). And, therefore, all search information can be employed in the ongoing computations—when searching for the next Pareto-optimal solution, GMAGS can start with optimization data from SOD instead of starting the computations from scratch. Within this approach, Rule 1 of the GMAGS algorithm can be formulated as follows:

Rule 1 (updated). Take the iteration points x_i, $0 \le i \le k+1$, and the scalar criterion values z_i, $0 \le i \le k+1$, from SOD (15).

(as can be seen, the iteration points in SOD are placed in order of increasing coordinate values).

In general, reusing search information requires fewer and fewer computations for solving each successive optimization problem down to executing just a few iterations to find the next efficient solutions.

5 Results of Computational Experiments

In this section, the results of computational experiments are presented. In the beginning, the ability to apply other optimization methods to solving multicriterial optimization problems was estimated. It is worth noting that the comparison of the methods needed to estimate this possibility should be performed on the optimization problems, the solution upon which the considered approach is oriented (the criteria can be multiextremal and computationally expensive—see Sect. 2). As a result, within the framework of the comparison performed, the multicriterial optimization methods, which are oriented to simpler MCO problem statements (for example, with the linear optimality criteria) may not be considered. It should also be stressed that in this paper, the approach based on scalarizing the vector criterion (see Sect. 3) is applied. Within this approach, each scalar problem (3) is a multiextremal optimization. Thus, the problem of comparing multicriterial optimization methods can be reduced to comparing the global search algorithms. Such comparisons have been reported in a sufficiently large number of publications—see, for example [1, 11–15, 26].

The experiments were organized as follows. First of all, a comparison of the GMAGS algorithm with the DIRECT method [10], which is a widely used global optimization method, was performed on a large set of multiextremal optimization problems. Experimental results have shown that GMAGS is more efficient. Based on these results and on the results of comparing various multiextremal optimization methods [1, 11–15, 26] in other experiments, the efficiency evaluation has only been performed for GMAGS. First, solving bi-criteria univariate MCO problems was performed. Then, the experiments for bi-criteria two-dimensional optimization problems were performed. Finally, experiments for a multicriterial optimization problem with 10 criteria were conducted.

In the first series of experiments, GMAGS is compared with the DIRECT method [10], which is a well-known global optimization method. The set of the test optimization functions consists of the multiextremal functions defined by relationship [14]:

$$\begin{gathered} AB = \left(\sum_{i=1}^{7} \sum_{j=1}^{7} [A_{ij} a_{ij}(y_1, y_2) + B_{ij} b_{ij}(y_1, y_2)] \right)^2 \\ CD = \left(\sum_{i=1}^{7} \sum_{j=1}^{7} [C_{ij} a_{ij}(y_1, y_2) - D_{ij} b_{ij}(y_1, y_2)] \right)^2 \\ \phi(y_1, y_2) = -\{AB + CD\}^{1/2} \end{gathered} \tag{16}$$

where

$$\begin{aligned} a_{ij}(y_1, y_2) &= \sin(\pi i y_1) \sin(\pi j y_2), \\ b_{ij}(y_1, y_2) &= \cos(\pi i y_1) \cos(\pi j y_2) \end{aligned}$$

are defined in the domain $0 \leq y_1, y_2 \leq 1$, and the parameters $-1 \leq A_{ij}, B_{ij}, C_{ij}, D_{ij} \leq 1$ are independent random numbers distributed uniformly. The minimization of such functions arises, for example, in the problem for estimating the maximum strain on a thin plate (determining its strength) at the transversal loading. The contour plots of two functions from this family are shown in Fig. 1—one can see that these types of function are essentially multiextremal.

In order to draw more substantiated conclusions on the efficiency of the compared methods, a set of 100 multiextremal problems has been solved.

For GMAGS the reliability parameter $r = 3$ and the search accuracy $\varepsilon = 0.01$ were used. All 100 problems have been solved with the required accuracy and the average number of executed optimization iterations is 512. For DIRECT, a search accuracy $\varepsilon = 10^{-6}$ has been used. In this case 91 problems have been solved and the average number of executed optimization iteration is 688. With a lower accuracy $\varepsilon = 10^{-2}$ the number of problems solved was 37.

Table 1 Experimental results for univariate MCO problems both with and without search information

Number of convolution coefficient values	Computations without using search information		Computations using search information		Reduction in the number of optimization iterations
	Total	Average per problem	Total	Average per problem	
1	34	34	34	34	**1**
10	497	49.7	153	15.3	**3.2**
20	1087	54.35	171	8.55	**6.4**
30	1578	52.6	174	5.8	**9.1**
40	2466	61.65	176	4.4	**14**
50	3363	67.26	178	3.56	**18.9**

Fig. 1 Contour plots of two multiextremal functions from the test optimization problem family

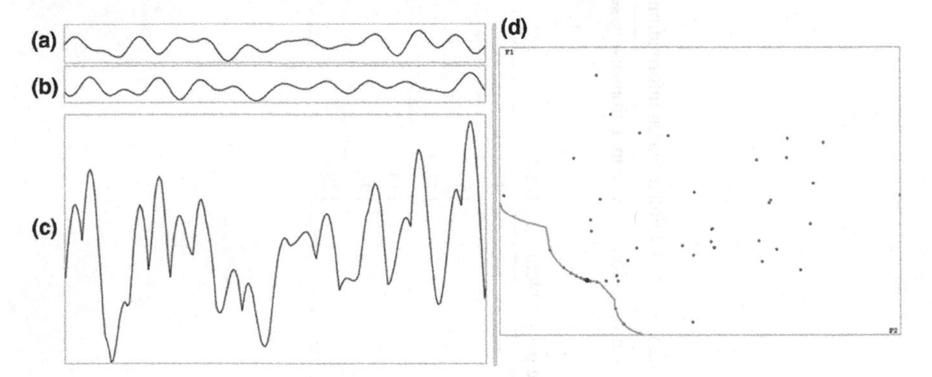

Fig. 2 An example of the test MCO problem: the criteria (**a, b**), the minimax convolution for $\lambda_1 = \lambda_2 = 0.5$ (**c**), the Pareto domain and the criteria values at the executed optimization iteration points (**d**)

Additional results from numerical comparisons of the GMAGS and DIRECT methods were also presented in [26].

In the next series of experiments, bi-criteria univariate MCO problems, i.e., for $N = 1$, $s = 2$ were solved. Like the criteria for the problems, the multiextremal functions are defined by the relationships [16]:

$$\phi(x) = A_0 + \sum_{i=1}^{14} (A_i \sin(2i\pi x) + B_i \cos(2i\pi x)), 0 \leq x \leq 1$$

where the coefficients A_0, A_i, B_i, $1 \leq i \leq 14$ were generated by a random number generator within the interval $[-1, 1]$. A graph of the function from such MCO problems is presented in Fig. 2: the criteria and the minimax convolution for

$\lambda_1 = \lambda_2 = 0.5$ are shown in the left panel, and a view of the Pareto domain and the criteria values at the executed optimization iteration points are shown in the right panel (the ordinate axis corresponds to the values of the first criterion, the abscissa axis—to the values of the second criterion).

To optimize the criteria convolution (5), GMAGS presented in Sect. 4 with the reliability parameter $r = 2$ and the search accuracy $\varepsilon = 0.001$ has been used. In each experiment, solving the problem (5) at several values of the convolution coefficients λ (from 1 up to 50 different values) has been performed both with search information and without it. The results are presented in Table 1.

The results presented here demonstrate that using search information in the experiments allowed the number of optimization iterations to be reduced in the solving multicriterial optimization problems by 3.2–18.9 times. In addition, as previously mentioned, solving each subsequent scalar problem (5) in order to find the next particular solution for the MCO problems requires performing fewer and fewer optimization iterations. For a clear presentation of this key property of the developed approach, the results of experiments are shown in Table 2 separately for the sequence of groups of solved problems (5); each group includes the scalar problems (5) for 10 values of the convolution coefficients λ (in total, 50 coefficients were selected for solving the MCO test problem).

The results of the experiments presented in Table 2 demonstrate a considerable reduction in the number of optimization iterations as the amount of obtained information increases. Thus, when solving a group of 10 scalar problems (5), starting from 21 convolutions of the partial criteria, only 2–3 optimization iterations are required until the termination condition is sufficiently fulfilled. The number of optimization iterations for the group of 41–50 problems was reduced by 448.5 times.

In the following series, experiments were performed to solve bi-criteria two-dimensional MCO problems, i.e., for $N = 2$, $s = 2$. The problems were formed using the multiextremal functions from (16) according to the rule

$$f_i(y) = (f_{1i}(y), f_{2i}(y)), 1 \le i \le 100,$$

where i, $1 \le i \le 100$ is the serial number of the multicriterial problem, and

$$f_{1i}(y) = \phi_i(y)$$

$$f_{2i}(y) = \begin{cases} \phi_{i+50}(y), & 1 \le i \le 50 \\ \phi_{i-50}(y), & 51 \le i \le 100 \end{cases}$$

(the lower indices at the functions $\phi_i(y)$ indicate the serial numbers for the multiextremal functions within the family of optimization problems).

The evaluation of the set of the efficient solutions (the Pareto set) necessary to conduct computational experiments has been performed numerically for each multicriterial problem being solved by means of scanning all the nodes of a uniform (within the search domain D) grid with a step of 10^{-3} in each coordinate (i.e., the grid had 1 million nodes total).

Table 2 The results of experiments for univariate MCO problems shown separately for the groups of 10 problems each

Number of problems included in a group	Computations without using search information		Computations using search information		Reduction in the number of optimization iterations
	Total	Average per problem	Total	Average per problem	
1–10	497	49.7	153	15.3	**3.2**
11–20	590	59	18	1.8	**32.8**
21–30	491	49.1	3	0.3	**163.7**
31–40	888	88.8	2	0.2	**444**
41–50	897	89.7	2	0.2	**448.5**

For a fuller representation of the efficiency of the developed approach, the GMAGS operational characteristics (see [14, 30]) have been constructed according to the results of the experiments performed. The operational characteristic (OC) of a method is a curve demonstrating the dependence of the number of solved problems from a certain class (the ordinate axis) on the number of trials (the abscissa axis) and is a set of pairs:

$$OC = \{(k, p(k)) : 1 \leq k \leq K\},$$

where k is the number of the optimization iterations, $p(k)$ is the fraction of the test class problems successfully solved within a given number of iterations, and K is the total number of the executed iterations. These indicators can be calculated based on the results of numerical experiments and are shown graphically in the form of a piecewise line graph. In general, one can examine the operational characteristic to show the probability of finding the global minimum with the required accuracy subject to the number of optimization iterations performed by the method.

The termination condition for GMAGS when solving the scalar problem (5) was the falling of the global optimization iteration point into the δ-nearness of any efficient solution from Pareto domain i.e.

$$\|y^k - y^*\| < \delta.$$

When carrying out the computational experiments, the following parameter values were used: the reliability parameter $r = 3$, the search accuracy $\varepsilon = 0.001$, and the size of δ-nearness $\delta = 0.001$. When solving each multicriterial optimization problem, a search has been performed for the efficient solutions to 50 various criteria convolution coefficients distributed uniformly.

The operational characteristics calculated based on the experimental results are presented in Fig. 3. The dashed line in Fig. 3 corresponds to solving the problems without reusing the search information while the solid line corresponds to reusing the search information.

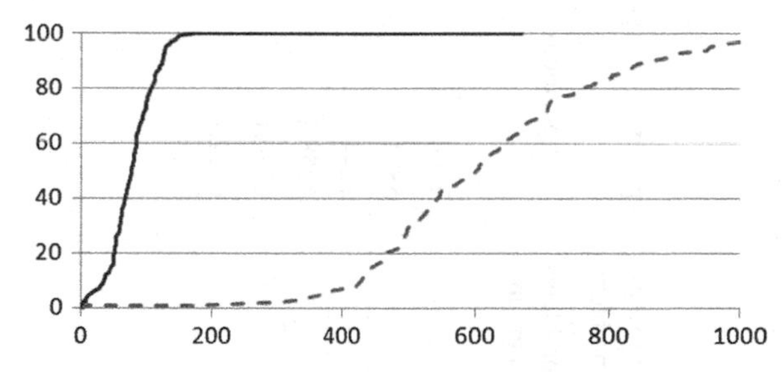

Fig. 3 Operational characteristics of GMAGS when solving multicriterial optimization problems without using the search information (*dashed line*) and with the search information (*solid line*)

Table 3 The results of experiments for groups of 10 problems separately (the data have been averaged over the results of solving 100 multicriterial two-dimensional problems)

Number of problems included in a group	Computations without using search information		Computations using search information		Reduction in the number of optimization iterations
	Total	Average per problem	Total	Average per problem	
1–10	5821.2	582.1	1232	123.2	**4.7**
11–20	7244.4	724.4	582.6	58.3	**12.4**
21–30	5258.2	525.8	512.7	51.3	**10.3**
31–40	6736.1	673.6	876.8	87.7	**7.7**
41–50	6112.1	611.2	779.5	78.0	**7.8**

As one can see, the results of this series of computational experiments confirm a significant reduction in the number of optimization iterations for solving multicriterial optimization problems using the search information obtained during the course of computations. Thus, within the execution of about 150 global optimization iterations without reusing search information, no problems were solved. The use of the search information has allowed practically all problems to be solved.

Similar to the data presentation on solving one-dimensional problems (Table 2), the computational results are shown separately for the sequence of the groups of solved problems (5); 10 scalar problems (5) have been included in each group.

According to the results presented above, the use of the search information in this series of experiments reduced the amount of optimization iterations performed up to 7.8 times (Table 3).

In the final experiment, the solution of the two-dimensional MCO problem with 10 criteria, i.e., for $N = 2, s = 10$, has been determined. The problems were formed using the multiextremal functions from (16). The following parameter values were used: the reliability parameter $r = 3.4$, search accuracy $\varepsilon = 0.005$, and the size of δ-nearness $\delta = 0.01$. As in previous experiments, the MCO problem (5) are solved using 50 various values of the convolution coefficients (49 problems have been solved with the required accuracy). The executed experiments demonstrated that the average number of optimization iterations for solving a single problem (5) without using the accumulated search information is 1813. In the case when GMAGS is taking into account the search information, the average number of optimization iterations is 305 (more than a fivefold reduction).

6 Conclusions

In this paper, an efficient approach is proposed for solving complex multicriterial optimization problems, where the optimality criteria may be multiextremal, and computing the criteria values may require a large volume of computations. The basis of the proposed approach consists of reducing multicriterial problems to nonlinear programming problems using the minimax convolution of the partial criteria, dimensionality reduction using Peano evolvents, and applying the efficient information-statistical methods of global optimization.

The key aspect of the approach consists in overcoming the high computational complexity of the global search for multiple efficient solutions in solving multicriterial optimization problems. A considerable enhancement in the efficiency and significant reduction in the volume of computations was achieved by maximizing the use of all search information obtained during the course of computations. For this purpose, it was necessary to provide the ability to store a large amount of search information, efficient processing, and using the search data during the course of solving multicriterial optimization problems. Within the framework of the developed approach, methods have been proposed for reducing all available search information to the values of current scalar nonlinear programming problem being solved. The search information reduced to the current state is used by the optimization methods for adaptive planning of the global search iterations to be performed.

According to the results of the computational experiments, this approach significantly reduces the computational costs for solving multicriterial optimization problems—by tens and hundreds times.

In conclusion, it is worth noting that the approach developed is promising and requires further investigation. First of all, it is necessary to continue conducting computational experiments to solve multicriterial optimization problems for a larger number of partial criteria of efficiency and for greater dimensionality of the optimization problems to be solved. Also, it is necessary to estimate the ability to organize parallel computations because of the high computation costs of solving global optimization problems.

Acknowledgements This work has been supported by Russian Science Foundation, project No 16-11-10150 "Novel efficient methods and software tools for time-consuming decision making problems using supercomputers of superior performance."

References

1. Barkalov, K.A., Gergel, V.P.: Multilevel scheme of dimensionality reduction for parallel global search algorithms. In: Proceedings of the 1st International Conference on Engineering and Applied Sciences Optimization, pp. 2111–2124 (2014)
2. Branke, J., Deb, K., Miettinen, K., Slowinski, R. (eds.): Multi-Objective Optimization—Interactive and Evolutionary Approaches. Springer, Berlin (2008)
3. Collette, Y., Siarry, P.: Multiobjective Optimization: Principles and Case Studies (Decision Engineering). Springer, Berlin (2011)
4. Deb, K.: Multi-Objective Optimization using Evolutionary Algorithms. Wiley, Chichester (2001)
5. Ehrgott, M.: Multicriteria Optimization. Springer (2005). (2nd edn., 2010)
6. Eichfelder, G.: Scalarizations for adaptively solving multi-objective optimization problems. Comput. Optim. Appl. **44**, 249–273 (2009)
7. Evtushenko, Y.G., Posypkin, M.A.: A deterministic algorithm for global multi-objective optimization. Optim. Methods Softw **29**(5), 1005–1019 (2014)
8. Figueira, J., Greco, S., Ehrgott, M. (eds.): Multiple Criteria Decision Analysis: State of the Art Surveys. Springer, New York (2005)
9. Floudas, C.A., Pardalos, M.P.: Recent Advances in Global Optimization. Princeton University Press (2016)
10. Gablonsky, J.M., Kelley, C.T.: A locally-biased form of the DIRECT algorithm. J. Glob. Optim. **21**(1), 27–37 (2001)
11. Gergel, V.P., Grishagin, V.A., Gergel, A.V.: Adaptive nested optimization scheme for multidimensional global search. J. Glob. Optim. **66**(1), 1–17 (2015)
12. Gergel, V., Grishagin, V., Israfilov, R.: Local tuning in nested scheme of global optimization. Procedia Comput. Sci. **51**, 865–874 (2015)
13. Gergel, V.P., Strongin, R.G.: Parallel computing for globally optimal decision making on cluster systems. Futur. Gener. Comput. Syst. **21**(5), 673–678 (2005)
14. Grishagin, V.A.: Operating characteristics of some global search algorithms. Probl. Stoch. Search **7**, 198–206 (1978). (In Russian)
15. Grishagin, V.A., Sergeyev, Y.D., Strongin, R.G.: Parallel characteristic algorithms for solving problems of global optimization. J. Glob. Optim. **10**, 185–206 (1997)

16. Hill, J.D.: A search technique for multimodal surfaces. IEE Trans. Syst. Cybern. **5**(1), 2–8 (1969)
17. Hillermeier, C., Jahn, J.: Multiobjective optimization: survey of methods and industrial applications. Surv. Math. Ind. **11**, 1–42 (2005)
18. Horst, R., Tuy, H.: Global Optimization: Deterministic Approaches. Springer, Berlin (1990)
19. Locatelli, M., Schoen, F.: Global Optimization: Theory, Algorithms, and Applications. SIAM (2013)
20. Koksalan, M.M., Wallenius, J., Zionts, S.: Multiple Criteria Decision Making: From Early History to the 21st Century. World Scientific, Singapore (2011)
21. Marler, R.T., Arora, J.S.: Survey of multi-objective optimization methods for engineering. Struct. Multidiscip. Optim. **26**, 369–395 (2004)
22. Marler, R.T., Arora, J.S.: Multi-Objective Optimization: Concepts and Methods for Engineering. VDM Verlag (2009)
23. Mardani, A., Jusoh, A., Nor, K., Khalifah, Z., Zakwan, N., Valipour, A.: Multiple criteria decision-making techniques and their applications—a review of the literature from 2000 to 2014. Econ. Res. Ekonomska Istraživanja **28**(1), 516–571 (2015). doi:10.1080/1331677X.2015.1075139
24. Miettinen, K.: Nonlinear Multiobjective Optimization. Springer (1999)
25. Pintér, J.D.: Global Optimization in Action (continuous and Lipschitz optimization: algorithms, implementations and applications). Kluwer Academic Publishers, Dortrecht
26. Sergeyev, Y.D., Kvasov, D.E.: Global search based on efficient diagonal partitions and a set of Lipschitz constants. SIAM J. Optim. **16**(3), 910–937 (2006)
27. Sergeyev, Y.D., Strongin, R.G., Lera, D.: Introduction to Global Optimization Exploiting Space Filling Curves. Springer (2013)
28. Siwale, I.: Practical multi-objective programming. In: Technical Report RD-14-2013. Apex Research Limited
29. Strongin, R.G.: Numerical Methods in Multiextremal Problems: Information-statistical Algorithms. Nauka, Moscow (1978). (in Russian)
30. Strongin, R., Sergeyev, Ya.: Global Optimization with Non-convex Constraints. Sequential and Parallel Algorithms. Kluwer Academic Publishers, Dordrecht (2000)
31. Tan, K.C., Khor, E.F., Lee, T.H.: Multi-objective Evolutionary Algorithms and Applications. Springer, London (2005)
32. Törn, A., Žilinskas, A.: Global Optimization. In: Lecture Notes in Computer Science, vol. 350. Springer, Berlin (1989)
33. Yang, X.-S.: Nature-Inspired Metaheuristic Algorithms. Luniver Press, Frome (2008)
34. Zavadskas, E.K., Turskis, Z., Kildiene, S.: State of art surveys of overviews on MCDM/MADM methods. Technol. Econ. Dev. Econ. **20**, 165–179 (2014)
35. Zhigljavsky, A.A.: Theory of Global Random Search. Kluwer Academic Publishers, Dordrecht (1991)
36. Žilinskas, A., Törn, A., Žilinskas, J.: Adaptation of a one-step worst-case optimal univariate algorithm of bi-objective Lipschitz optimization to multidimensional problems. Commun. Nonlinear Sci. Numer. Simul. **21**, 89–98 (2015)

Comparison of Two Heuristic Algorithms for a Location and Design Problem

Alexander Gnusarev

Abstract The article is devoted to the decision-making methods for the following location and design problem. The Company is planning to locate its facilities and gain profit at an already existing market. Therefore it has to take into consideration such circumstances as already placed competing facilities; the presence of several projects for each facility opening; the share of the served demand is flexible and depends on the facility location. The aim of the Company is to determine its new facilities locations and options in order to attract the biggest share of the demand. Modeling flexible demand requires exploiting nonlinear functions which complicates the development of the solution methods. A Variable Neighborhoods Search algorithm and a Greedy Weight Heuristic are proposed. The experimental analysis of the algorithms for the instances of special structure has been carried out. New best known solutions have been found, thus denoting the perspective of the further research in this direction.

1 Introduction

Among the numerous applications there is a great interest in location problems [1]. In many cases, it is necessary to place the facilities in several locations and to assign a number of customers to be served to them so, that the total expenses were the least. In the well-known classical models, such as Simple Plant Location Problem, p-median Problem, and others [1]; the decision is made by one person, not taking into consideration the customers' opinions and other circumstances.

However, the customers very often have their own preferences, and several rival companies struggle for serving them. Such situations are described in competitive market location problems which most accurately characterize the actual business environment. The paper deals with a discrete type of such a task in which the facilities can be opened at a finite set of possible locations and the clients are in a discrete set of points. The rival companies have already occupied some of them and are unable to change their decision. The new Company is aware of the existing competition. The

A. Gnusarev (✉)
Sobolev Institute of Mathematics SB RAS, Omsk Branch, Pevtsova str. 13, Omsk 644043, Russia
e-mail: alexander.gnussarev@gmail.com

V.A. Kalyagin et al. (eds.), *Models, Algorithms, and Technologies for Network Analysis*, Springer Proceedings in Mathematics & Statistics 197,
DOI 10.1007/978-3-319-56829-4_4

customers choose the facilities taking into account their attractiveness. This task is related to the Static Probabilistic Competitive Facility Location Problems [2] one of them was develosped by Berman and Krass [3].

We take into consideration a special case of the model in [3] which was proposed in [4] and is called "facility location and design problem". Aboolian et al. proposed a spatial interaction model for describing demand cannibalization. In [4], the demand is a function of the total utility and the objective function is nonlinear. This location problem is quite complicated both from the theoretical and the practical points of view. Obtaining optimal solutions for large instances of the problem using the exact algorithms, including software packages, can require significant time and computer resources. In [4], the adapted weighted greedy heuristic algorithm is proposed for the solution of the discrete competitive facility location and design problem. Therefore, it is interesting to develop modern decision methods for the considered problem.

A lot of attention has been given to the methods of finding approximate solutions recently [5], they include the class of local search algorithms [6]. Variable Neighborhood Search Algorithm (VNS) belongs to this class, it is successfully used for solution of many applied tasks. The basic idea of VNS is to explore a set of predefined neighborhoods successively in order to provide a better solution. The Variable Neighborhood Search Approach [7] for the location and design problem is developed in this paper. A new version of the VNS is proposed, the specific types of neighborhoods are described. The numerical experiments based on the specially generated instances [8] have been carried out, the comparison of Variable Neighborhood Search Algorithm and Greedy Weight Heuristic [4] have been executed. CoinBonmin is used for the analysis of the quality of the obtained solutions [9]. The results show that VNS can find new best known solutions to the large instances of the problem with a small relative error.

The rest of the paper is organized as follows. Section 2 contains the formulation of the problem. Section 3 describes the schemes of the Neighborhood Search Algorithm and Greedy Weight Heuristic. The new version of VNS for the considered problem is proposed in Sect. 3 as well. Section 4 presents the results of the numerical experiments. Section 5 concludes the paper.

2 Problem Formulation

The article deals with the situation when a new Company plans to enter the market of existing products and services. It makes a decision to open a supermarket chain, which will differ in size or the set of goods provided from the existing ones. Such differences are called "design options". Customers select companies based on their attractiveness and distance from their location. The aim of the Company is to interest the greatest number of the customers thus serving the largest share of the demand. This percentage is not fixed for the company and depends on where and which option the new enterprise will be opened on.

The problem has several applications. One of them appeared in Toronto (Canada) and is described by Kraas, Berman, Aboolian. The mathematical model has been formulated in [4]. Let R be the set of facility designs, $r \in R$. There are w_i customers at the point i of discrete set $N = \{1, 2, \cdots n\}$ in the problem. All the customers have the same demands, so each item can be considered as one client with the weight w_i. Let the distance d_{ij} between the points i and j be measured in Euclidean metric or it equals to the shortest part leght in the corresponding graph. Let $P \subseteq N$ be the set of potential facility locations. It is assumed that $C \subset P$ is the set of preexisting rival facilities. The Company may open its branches in $S = P \setminus C$ taking into account the available budget B, attractiveness a_{jr}, and the cost of opening c_{jr} facility $j \in S$ with design $r \in R$.

Such flexible choice of customer is represented in the gravity-type spatial interaction models. These models are known as the brand share models in the marketing literature [10]. According to these models, the utility u_{ij} of a facility at location $j \in S$ for a customer at point $i \in N$ can be written as exponential function. Let $x_{jr} = 1$, if the facility j is opened with the design variant r and $x_{jr} = 0$ otherwise, $j \in S, r \in R$.

To determine the usefulness u_{ij} of the facility $j \in S$ for the customer $i \in N$ the coefficients k_{ijk} have been introduced: $k_{ijk} = a_{jr}(d_{ij} + 1)^{-\beta}$. They depend on the sensitivity β of the customers to the distance from the facility.

The utility $u_{ij} = \sum_{r=1}^{R} k_{ijr}x_{jr}$. The total utility of the customer $i \in N$ received from the competitive facilities is

$$U_i(C) = \sum_{j \in C} u_{ij}.$$

The demand function is

$$g(U_i) = 1 - \exp\Big(- \lambda_i U_i \Big),$$

where λ_i is the flexible demand characteristic in point i; U_i is the total utility for a customer at $i \in N$ from all open facilities:

$$U_i = \sum_{j \in S} \sum_{r=1}^{R} k_{ijr}x_{jr} + U_i(C).$$

The total share of the company in the facility $i \in N$:

$$MS_i = \frac{U_i(S)}{U_i(S) + U_i(C)} = \frac{\sum_{j \in S} \sum_{r=1}^{R} k_{ijr}x_{jr}}{\sum_{j \in S} \sum_{r=1}^{R} k_{ijr}x_{jr} + \sum_{j \in C} u_{ij}}.$$

The mathematical model looks like:

$$\sum_{i \in N} w_i \cdot g(U_i) \cdot MS_i \rightarrow max \tag{1}$$

$$\sum_{j \in S} \sum_{r=1}^{R} c_{jr} x_{jr} \leq B, \tag{2}$$

$$\sum_{r=1}^{R} x_{jr} \leq 1, \tag{3}$$

$$x_{jr} \in \{0, 1\}, \quad r \in R, j \in S. \tag{4}$$

Based on the above notation, the objective function (1) looks as

$$\sum_{i \in N} w_i \left(1 - \exp\left(-\lambda_i \left(\sum_{j \in S} \sum_{r=1}^{R} k_{ijr} x_{jr} + U_i(C)\right)\right)\right) \cdot \tag{5}$$

$$\cdot \left(\frac{\sum_{j \in S} \sum_{r=1}^{R} k_{ijr} x_{jr}}{\sum_{j \in S} \sum_{r=1}^{R} k_{ijr} x_{jr} + \sum_{j \in C} u_{ij}} \right) \rightarrow max.$$

The objective function (5) reflects the Company's aim to maximize the volume of the customers' demand. Inequality (2) takes into account the available budget. Condition (3) shows that only one variant of the design can be selected for each facility.

3 Algorithms

The of well-known software GAMS (CoinBonmin) for the location and design problem can calculate an approximate solution which is not guaranteed to be optimal [9]. Solving such problem requires a significant investment of time and computing resources. In this regard, one of the approaches to its solution is in employment of the approximate heuristic methods. In this paper, the Variable Neighborhoods Search algorithm [7, 11] has been constructed and compared to the Greedy Weight Heuristic one for the considered problem, the scheme of the Variable Neighborhood Search algorithm (VNS) is given below.

The scheme of the VNS algorithm

Initialization. Select the set of neighborhood structures N_k, $k = 1, \ldots, k_{max}$, that will be used in the search; find the initial solution x; choose the stopping condition.

Repeat the following steps until the stopping condition is met.

(1) Set $k := 1$.
(2) Until $k = k_{max}$, repeat the following steps.

(a) *Shaking.* Generate a point $x\prime$ at random from the k-th neighborhood of x ($x\prime \in N_k(x)$);
(b) *Local search.* Apply a local search method with $x\prime$ as the initial solution; denote $x\prime\prime$ as the obtained local optimum;
(c) *Move or not.* If this local optimum is better than the incumbent, move there $x := x\prime\prime$, and continue the search with $N_1, k := 1$; otherwise, set $k := k + 1$.

The new types of neighborhoods used for the algorithm are described below. Let the vector $z = (z_i)$ be such that z_i corresponds to facility i: $z_i = r$ iff $x_{ir} = 1$. The feasible initial solution z is obtained using a special deterministic procedure.
Neighborhood 1 (N1). Feasible solution z' is called neighboring for z if it can be obtained with the following steps:

(a) choose randomly one of the open facilities p with the scenario z_p and close it;
(b) select the facility q which is closed and has the highest attractiveness; then open the facility q with the scenario z_p.

Neighborhood 2 (N2). Feasible solution z' is called neighboring for z if it can be obtained with the following operations:

(a) choose randomly one of the open facilities p with the scenario z_p and reduce the number of the scenario;
(b) select randomly the facility q and increase the number of its scenario.

Neighborhood 3 (N3). Unlike the Neighborhood 2 at step (b) select randomly the facility q which is closed; then open the facility q with the scenario z_p.
Lin–Kernighan neighborhood was applied as *Neighborhood 4* [12].
In order to describe Greedy Weight Heuristic let us introduce the following terms: L is a set of facility design pairs (j, r), where $j \in S$ is the location of the facility, $r \in R$ is the design scenario chosen for that location. $Z(L)$ is the objective function value associated with the location-design set L. Let $\rho_{jr}(L) = Z(L \cup (j, r)) - Z(L)$ be the improvement of the objective function obtained by adding the location-design pair (j, r) to the location-design set L. Set T is the location-design pairs that should be excluded from further consideration. The outline of Greedy Weight Heuristic is as follows:
Step 1: $L^0 = \emptyset, T^0 = \emptyset, t = 1$
Step 2: Let $(j(t), r(t)) = arg\ max_{(j,r)\notin T^{t-1}}\{\frac{\rho_{jr}(L^{t-1})}{c_{jr}}\}$
If $\sum_{(j,r)\in L^{t-1}} c_{jr} + c_{j(t)r(t)} \leq B$ then
set $L^t = L^{t-1} + \{(j(t), r(t))\}$, $T^t = T^{t-1} + \{(j(t), r)|r \in R\}$
and set $t = t + 1$.
Return to *Step 2*.
Else go to *Step 3*.
Endif

Step 3: If $Z(L^{t-1}) \geq Z(j(t), r(t))$, then
$L^H = L^{t-1}$ is the adapted greedy solution with value $Z(L^H)$.
Else set $L^1 = \{(j(t), r(t))\}$, $T^1 = \{(j(t), r) | r \in R\}$, $t = 2$
Return to *Step 2*.
Endif
Stop

The described algorithms have been programmed on a computer and experimentally investigated. The results are contained in the following section.

4 Experimental Study

The validation of the VNS algorithm has been conducted for the following data: the neighborhoods N1, N2, N3 and Lin–Kernighan are used, local descent has been carried out with the help of the neighborhood of Lin–Kernighan with 9 points. Stopping criteria for the VNS has been the complete neighborhood exploration without an improvement of the solution.

There are three options for the enterprises development ($R = 3$): the small one with the cost of opening $c_{j1} = 1$; the medium one with the cost of opening $c_{j2} = 2$; the large one with the cost of opening $c_{j3} = 3$ for all $j \in S$. At each point of the

Table 1 Best known solutions

Tests	Arbitrary distances			Euclidean distances		
	GAMS	VNS	GWH	GAMS	VNS	GWH
300.3.1	36.15	36.143	23.190	—	35.183	35.183
300.3.2	54.12	57.158	44.283	—	54.446	54.446
300.3.3	74.41	75.513	70.392	—	73.053	73.053
300.3.4	94.47	96.097	86.411	—	91.081	91.081
300.5.1	30.02	30.334	21.419	—	30.899	30.899
300.5.2	51.04	51.051	38.392	—	50.514	50.515
300.5.3	66.99	67.503	54.510	—	69.360	69.360
300.5.4	81.93	86.312	70.427	—	87.756	87.848
300.7.1	36.43	36.427	20.095	—	36.154	36.154
300.7.2	53.69	55.627	39.798	—	57.568	57.568
300.7.3	74.62	74.610	58.637	—	77.670	77.787
300.7.4	92.83	95.266	76.566	—	96.992	97.119
300.9.1	31.83	31.823	20.322	—	32.093	32.093
300.9.2	48.65	51.274	38.876	—	51.503	51.503
300.9.3	67.09	69.000	56.537	—	70.294	70.295
300.9.4	85.87	85.603	73.210	—	88.947	88.947

Table 2 CPU time (sec)

Tests $\vert N\vert$	Arbitrary distances						Euclidean distances					
	VNS			GWH			VNS			GWH		
	min	av	max	min	av	max	min	av	max	min	av	max
60	10.94	21.92	46.16	0.06	0.12	0.19	12.29	20.52	39.26	0.10	0.19	0.36
80	21.32	34.16	86.34	0.14	0.23	0.32	23.04	36.49	83.92	0.29	0.46	0.63
100	38.01	60.95	152.16	0.25	0.52	0.99	32.58	48.47	100.74	0.43	0.72	1.14
150	76.11	97.48	141.28	0.54	1.08	2.61	75.41	145.97	451.60	1.51	2.91	4.23
200	225.05	183.43	295.40	1.65	2.66	4.55	109.51	222.46	447.04	3.44	5.65	8.71
300	265.45	438.38	643.23	3.96	9.14	13.75	268.04	614.21	1408.92	10.68	18.24	25.85

demand a business can be opened, i.e., $P = N$. The budget varies from 3 to 9 in the increments of 2 units. For example, having the budget of 9 units the Company can open 3 large enterprises or 9 small or it can combine them. The problem has been considered for random distances ($d_{ij} \in [0, 30]$, $i \neq j$, $d_{ij} = d_{ji}$) and satisfying the triangle inequality (coordinate $x \in [0, 100]$, coordinate $y \in [0, 150]$). The number of obtained locations is 60, 80, 100, 150, 200, 300. It has been assumed that the problem possesses a high sensitivity to the distance ($\beta = 2$) and a nonfixed demand, i.e., $\lambda = 1$.

Experiments were carried out on PC Intel i5–2450M, 2.50 GHz, 4GB memory. The test cases with Euclidean distances are proved to be difficult for the CoinBonmin solver. In particular, the maximum CPU time for the test problems with $|N| = 60$ was more than 63 h. Therefore, CoinBonmin was given 10 min of CPU time for each example of higher dimensions. Solver CoinBonmin found the best known solutions in 13 cases out of 80 (see Table 1). The VNS and GWH algorithms found new best known solutions for all the test problems with Euclidean distances with dimensions from 80 to 300 in less time. The average time of the VNS algorithm until the stopping criterion was triggered is 181.35 s. (GWH 2.29 s). In all the cases, the best solutions of the VNS were equal to the best solutions of GWH.

Table 2 contains the information about minimal (min), average (av), and maximal (max) CPU time (in seconds) of the proposed algorithms for the test problems. GAMS found records for all the tasks in the test cases with arbitrary distances. The average improvement of the VNS from GAMS was 1.55%, the average deviations of the GWH from GAMS was 12.5%. The results are presented in Figs. 1 and 2. The VNS algorithm improved the record values found by GWH in all test instances with arbitrary distances.

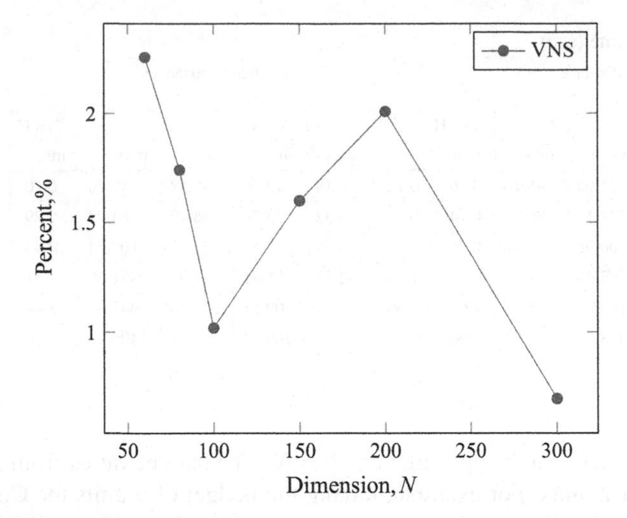

Fig. 1 The average improvement of new best known solutions obtaining by VNS upon CoinBonmin

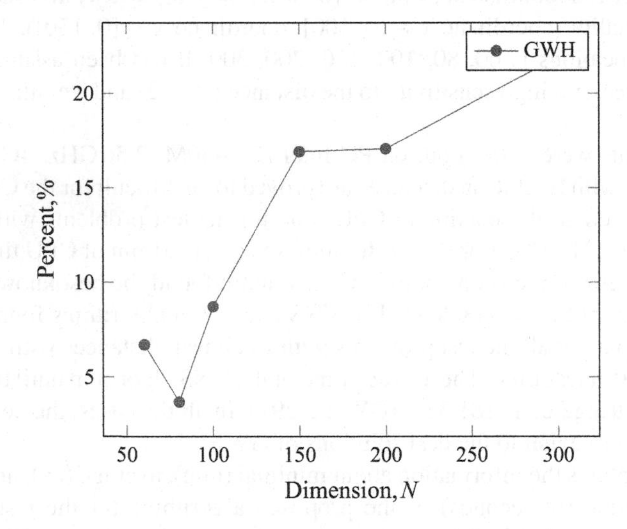

Fig. 2 The average deviation of GWH from the CoinBonmin results (It is information about how many percent on average GWH results worse than CoinBonmin results)

5 Conclusion

In this paper, we have created a new version of the Variable Neighborhood Search algorithm and implemented the Greedy Weight Heuristic for the location and design problem. New neighborhoods of a special type have been proposed, the experimental tuning of parameters for both algorithms have been carried out. Two sets of specially structured test examples have been generated. The proposed algorithms have been able to gain new best known solutions or solutions with small relative error. Therefore, the obtained results have indicated the usefulness of the proposed algorithms for solving the problem.

Acknowledgements This research was supported by the Russian Foundation for Basic Research, grant 15-07-01141.

References

1. Mirchandani, B.P., Francis, R.L. (eds.).: Discrete Location Theory. Wiley-Interscience, New York (1990)
2. Karakitsion, A.: Modeling Discrete Competitive Facility Location. Springer, Heidelberg (2015)
3. Berman, O., Krass, D.: Locating multiple competitive facilities: spatial interaction models with variable expenditures. Ann. Oper. Res. **111**, 197–225 (2002)
4. Aboolian, R., Berman, O., Krass, D.: Competitive facility location and design problem. Eur. J. Oper. Res. **182**, 40–62 (2007)
5. Gendreau, M., Potvin, J.Y.: Handbook of Metaheuristics. Springer, New York (2010)
6. Yannakakis, M.: Computational complexity. In: Aarts, E., Lenstra, J.K. (eds.) Local Search in Combinatorial Optimization, pp. 19–55. Wiley, Chichester (1997)
7. Hansen, P., Mladenovic, N.: Variable neighborhood search: principles and applications (invited review). Eur. J. Oper. Res. **130**(3), 449–467 (2001)
8. Levanova, T., Gnusarev, A.: Variable neighborhood search approach for the location and design problem. In: Kochetov, Y et al. (eds.) DOOR-2016. LNCS, vol. 9869, pp. 570–577. Springer, Heidelberg (2016)
9. Bonami, P., Biegler, L.T., Conn, A.R., Cornuejols, G., Grossmann, I.E., Laird, C.D., Lee, J., Lodi, A., Margot, F., Sawaya, N., Wachter, A.: An algorithmic framework for convex mixed integer nonlinear programs. Discrete Optimization **5**(2), 186–204 (2008)
10. Huff, D.L.: Defining and estimating a trade area. J Mark. **28**, 34–38 (1964)
11. Hansen, P., Mladenovic, N., Moreno-Perez, J.F.: Variable neighbourhood search: algorithms and applications. Ann. Oper. Res. **175**, 367–407 (2010)
12. Kochetov, Y., Alekseeva, E., Levanova, T., Loresh, M.: Large neighborhood local search for the p-median problem. Yugoslav J. Oper. Res. **15**(2), 53–64 (2005)

A Class of Smooth Modification of Space-Filling Curves for Global Optimization Problems

Alexey Goryachih

Abstract This work presents a class of smooth modifications of space-filling curves applied to global optimization problems. These modifications make the approximations of the Peano curves (evolvents) differentiable in all points, and save the differentiability of the optimized function. To evaluate the proposed approach, some results of numerical experiments with the original and modified evolvents for solving global optimization problems are discussed.

1 Introduction

Problems of global optimization are actively studied and can be often found in various applications [1–7]. A key feature of these problems is that several local optima can exist and finding the global optimum requires the analysis of the whole search domain. To provide this analysis, global optimization methods should process multidimensional search data obtained in the course of computations. Such computations can be extremely difficult even for low-dimensional problems and rise exponentially with increasing the dimension of optimization problems. As a result, many optimization algorithms, to various extents, use dimension reduction methods [4, 5, 8–10].

The structure of the paper is as follows. Section 2 describes the statement of the global optimization problem. Section 3 presents the discussion of issues of the dimension reduction methods based on the Peano curves. Section 4 introduces the developed techniques for construction of smooth modifications of these curves. Section 5 describes a generalized multidimensional algorithm of global search for solving the reduced multiextremal optimization problems. The results of numerical experiments

---

A. Goryachih (✉)
Lobachevsky State University of Nizhni Novgorod, 23, Gagarin Prospect, Nizhni Novgorod 603600, Russia
e-mail: goryachihalexeysergeevich@gmail.com

V.A. Kalyagin et al. (eds.), *Models, Algorithms, and Technologies for Network Analysis*, Springer Proceedings in Mathematics & Statistics 197,
DOI 10.1007/978-3-319-56829-4_5

are presented in Sect. 6. The discussions of the obtained results and basic directions for further research are given in the Conclusion, after which follows the Acknowledgments.

2 Problem Statement

The global optimization problem can be stated as a problem of finding the minimum value of a function $\varphi(y)$

$$\begin{aligned} \varphi(y^*) &= \min\{\varphi(y) : y \in D\}, \\ D &= \{y \in R^N : -a_i \le x_i \le b_i, 1 \le i \le N\}, \end{aligned} \tag{1}$$

where D is the search domain, and $a, b \in R^N$ are the given vectors.

It is supposed that the function $\varphi(y)$ satisfies the Lipschitz condition

$$|\varphi(y_2) - \varphi(y_1)| \le L \mid\mid y_2 - y_1 \mid\mid, y_1, y_2 \in D, \tag{2}$$

where $L > 0$ is the Lipschitz constant, and $\mid\mid * \mid\mid$ is norm in the space R^N.

Suppose that $\varphi(y)$ is differentiable for possible applicability optimization algorithms that use derivatives. For simplicity of further explanation also assume that D is a unit hypercube.

The numerical solving of the problem (1) is usually considered as the generation of a sequence of optimization points y_k which converges to the global optimum y^*, and the stop condition might be $\mid\mid y_k - y^* \mid\mid \le \varepsilon$, where $\varepsilon > 0$ is the given accuracy.

3 Dimension Reduction Based on the Peano Curves

Dimension reduction allows reducing the solving a multidimensional problem to solving a family of the one-dimensional ones and decreasing the complexity of the optimization algorithms.

One of the possible approaches of dimension reduction is based on employing the space-filling Peano curves, which map the unit interval [0, 1] onto the N-dimensional hypercube D (1) continuously. For example, this approach is widely used in the information statistical approach to the global optimization [5, 9, 11, 12].

The first space-filling curve was introduced by Giuseppe Peano in 1890. The space-filling curve, which is used for global optimization problems, was introduced by David Hilbert in 1891 (see Fig. 1). The computational schemes for these curves allow constructing an approximation to the Peano curve for any given accuracy (for simplicity of further explanation, such approximations will be named as the Peano or space-filling curves).

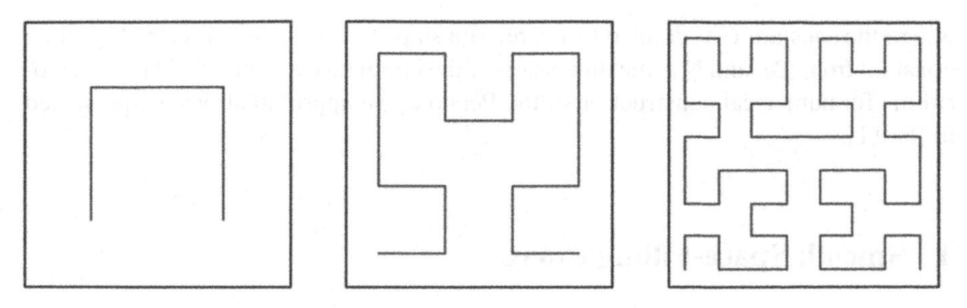

Fig. 1 The first three steps of construction of a Peano curve using the Hilbert scheme

The details of construction of the Peano curves using the Hilbert scheme are described in [5, 11]. It is necessary to indicate that the curve is constructed iteratively, and after m iterations the Peano curve will look like

$$y(x) = y_i + (y_{i+1} - y_i)\frac{x - x_i}{x_{i+1} - x_i}, x \in [x_i, x_{i+1}], 0 \le i \le 2^{Nm} - 1, \quad (3)$$

where x_i and y_i $(0 \le i \le 2^{Nm})$ are the uniform grids for the unit interval and the unit hypercube, correspondently, satisfying the relationships

$$\begin{aligned} |x_{i+1} - x_i| &= 2^{-Nm}, \\ \| y_{i+1} - y_i \| &= 2^{-m}, \\ 0 \le i &\le 2^{Nm} - 1, \end{aligned} \quad (4)$$

and also y_i, y_{i+1} have only one different coordinate.

Since the Peano curve $y(x)$ from (3) and $\varphi(y)$ from (1) and (2) are continuous, these ones satisfy the following relationship:

$$\varphi(y^*) = \varphi(y(x^*)) = \min\{\varphi(y(x)) : x \in [0, 1]\}. \quad (5)$$

Relationship (5) allows reducing the solving of the multidimensional problem (1) to solving the one-dimensional problem (5).

Such approach makes it possible to use a majority of the developed efficient one-dimensional global search algorithms for solving multidimensional optimization problems (it is necessary to note that applying such one-dimensional algorithms should go through correspondent adaptation), see for the examples [13–19]

It is also important to note that this dimension reduction maps the multidimensional problem (1) with the Lipschitz condition (2) onto the one-dimensional problem (5), in which the function $\varphi(y(x))$ satisfies the uniform Holder condition in the form (6)

$$|\varphi(y(x_2)) - \varphi(y(x_1))| \le G \| x_2 - x_1 \|^{1/N}, x_1, x_2 \in [0, 1], \quad (6)$$

where the constant G is defined by the relationship $G = 4L\sqrt{(N)}$, L is the Lipschitz constant from (2), and N is the dimension of the initial problem (1) [5, 11]. The algorithms for numerical construction of the Peano curve approximations are presented in [5, 11].

4 Smooth Space-Filling Curve

It is important to note that the smoothness lack of the Peano curve (3) leads to nondifferentiability of the function $\varphi(y(x))$. Therefore, the methods based on using derivatives cannot be adapted for solving the one-dimensional problem (5). This fact limits the possibility of increasing the efficiency of global search. For example, as it has been shown in [20–22], the use of differentiability can speed up the numerical solving of the global optimization problems more than 10–20 times in terms of the number of function value calculations. To address this issue, it is necessary to develop the modifications of the Peano curves, that can ensure smoothness of the reduced one-dimensional problems.

The smoothing of the Peano curves based on the replacement of the piecewise linear segments (see (3) and Fig. 1) by smooth arcs is suggested in this work. Several modifications of the Peano curves with different degree of smoothing will be proposed below.

In order to define the degree of smoothing, the parameter H will be used, it defines the relative size of smoothing area from each end of a segment $[y_i, y_i + 1], 0 \leq i \leq 2Nm - 1$ with respect to the length of the whole segment (see Fig. 2)

Any three successive points y_{i-1}, y_i, y_{i+1} from (3) can have no more than two different coordinates. If these ones have one different coordinate, then a smooth arc is not required. If the points have two different coordinates (for instance, let these coordinates be l_1 and l_2), then a construction of a curve smooth in the corresponding plane should be applied (see Fig. 4). A smooth arc can be defined by two polynomials (7) (for simplicity of explanation of the approach, the illustration was made for the planar case, i.e., for N = 2, Fig. 3).

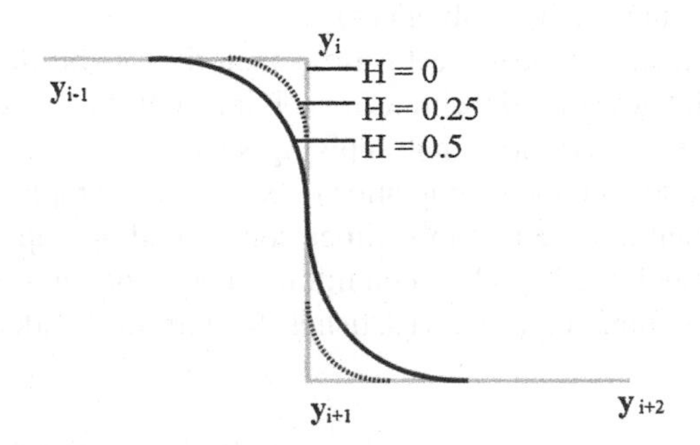

Fig. 2 The dependence of the size of the smoothing arcs on the parameter H

Fig. 3 A smooth modification of the Space-filling Curves

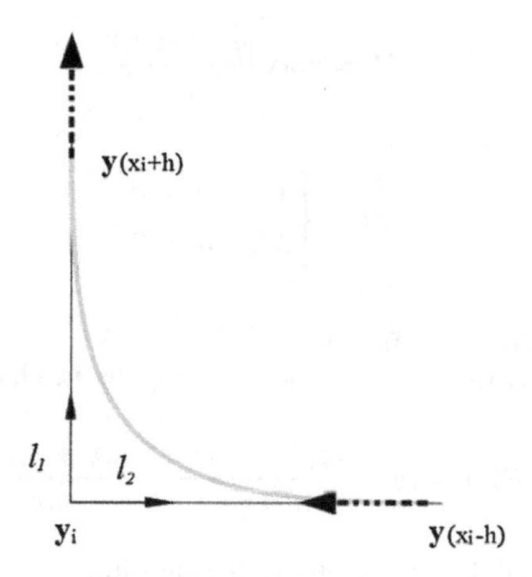

Fig. 4 The construction of a smooth arc

$$P(x) = f_0 + (x - x_0)\left\{f_0' + (x - x_0)[f_0' - (f_0 - f_1)/(x_0 - x_1) + \right.$$
$$\left. + (x - x_1)(f_0' - 2(f_0 - f_1)/(x_0 - x_1) + f_1')/(x_0 - x_1)]/(x_0 - x_1)\right\}, \quad (7)$$

where $x_0 \leq x \leq x_1$, x_0, and x_1 points on a unit interval ($x_0 = x_i - h$, $x_1 = x_i + h$, $h = H2^{-Nm}$), f_0, f_1 and f_0', f_1' are the coordinates l_1 and l_2 of the points and the derivatives of the Peano curve $y(x)$ at these points—see Fig. 4.

$$f_0 = y_{l_1/l_2}(x_0),\ f_1 = y_{l_1/l_2}(x_1),$$
$$f'_0 = y_{l_1/l_2}(x_0),\ f'_1 = y_{l_1/l_2}(x_1) \tag{8}$$

5 Optimization Method for Solving Reduced Optimization Problems

To solve the reduced the one-dimensional problems (5), well-known Generalized Multidimensional Algorithm of Global Search (GMAGS) proposed by Strongin [5, 12] was used. The computational scheme of the algorithm can be described as follows:

Starting optimization points x^1 and x^2 are fixed in such way that $x^1 = 0$, $x^2 = 1$, and number of trials k is set as 2. The point $x^{k+1}, k \geq 2$ of the next $(k+1)^{th}$ iteration is chosen as follows:

1. Renumber the optimization points $x^i, 1 \leq i \leq k$ of the previous iterations by subscripts so that
$$0 = x_0 < x_1 < \ldots < x_{k-1} = 1. \tag{9}$$
2. Compute the value m being an estimate of the Holder constant G from (6) as follows:
$$M = \max \frac{|\varphi_i - \varphi_{i-1}|}{\rho}, \tag{10}$$
and set
$$m = \begin{cases} 1, & M = 0 \\ M, & M > 0 \end{cases}, \tag{11}$$
where $\varphi_i = \varphi(y(x_i))$, $\rho_i = (x_i - x_{i-1})^{\frac{1}{N}}$, $1 \leq i \leq k+1$.
3. For each interval (x_i, x_{i-1}), $2 \leq i \leq k-1$ compute the characteristics
$$R(i) = \rho_i + \frac{(\varphi_i - \varphi_{i-1})^2}{m^2 r^2 \rho_i} - 2\frac{\varphi_i + \varphi_{i-1}}{mr}, \tag{12}$$
where $r > 1$ the reliability parameter of the algorithm.
4. Find the interval (x_t, x_{t-1}) that corresponds to the maximal characteristic
$$R(t) = \max\{R(i), 1 \leq i \leq k-1\}. \tag{13}$$
5. Execute the next iteration at the point
$$x^{k+1} = \frac{1}{2}(x_t + x_{t-1}) - sign(\varphi_t - \varphi_{t-1})\frac{1}{2r}\left[\frac{\varphi_t - \varphi_{t-1}}{m}\right]^N, \tag{14}$$
then calculate value $\varphi(y(x^{k+1}))$ and increase the number of iterations $k := k+1$.

Steps 1–5 of GMAGS are executed until $\rho_t \leq \varepsilon$, where $\varepsilon > 0$ is a given accuracy,

6 Results of Numerical Experiments

The numerical experiments were performed for evaluation of efficiency of the smooth modifications of the Peano curves.

Within the executed experiments a set of 100 multiextremal Grishagin's functions was chosen [5, 11, 12, 23]. These functions are defined according to relationships

$$\varepsilon(y_1, y_2) = -\left\{\left(\sum_{i=1}^{7}\sum_{j=1}^{7}\left[A_{ij}a_{ij}(y_1, y_2) + B_{ij}b_{ij}(y_1, y_2)\right]\right)^2 + \right.$$

$$\left. + \left(\sum_{i=1}^{7}\sum_{j=1}^{7}\left[C_{ij}a_{ij}(y_1, y_2) + D_{ij}b_{ij}(y_1, y_2)\right]\right)^2\right\}^{\frac{1}{2}}, \tag{15}$$

$$a_{ij}(y_1, y_2) = \sin(\pi i y_1)\sin(\pi j y_2),$$
$$b_{ij}(y_1, y_2) = \cos(\pi i y_1)\cos(\pi j y_2),$$

where $0 \leq y_1, y_2 \leq 1, -1 \leq A_{ij}, B_{ij}, C_{ij}, D_{ij} \leq 1$ are the independent random generated parameters. Parameter of GMAGS $r = 3.0$, accuracy $\varepsilon = 10^{-3}$. The numerical experiments were performed for the Peano curve ($H = 0$) and its smooth modification with the parameter $H = 0.1, 0.25, 0.4$, and 0.5 correspondingly.

As an example, the optimization points for solving a test problem from (15) using the Peano curve and its smooth modification ($H = 0.5$) are shown in Fig. 5. This result shows a significant advantage of the smooth modification in some particular cases.

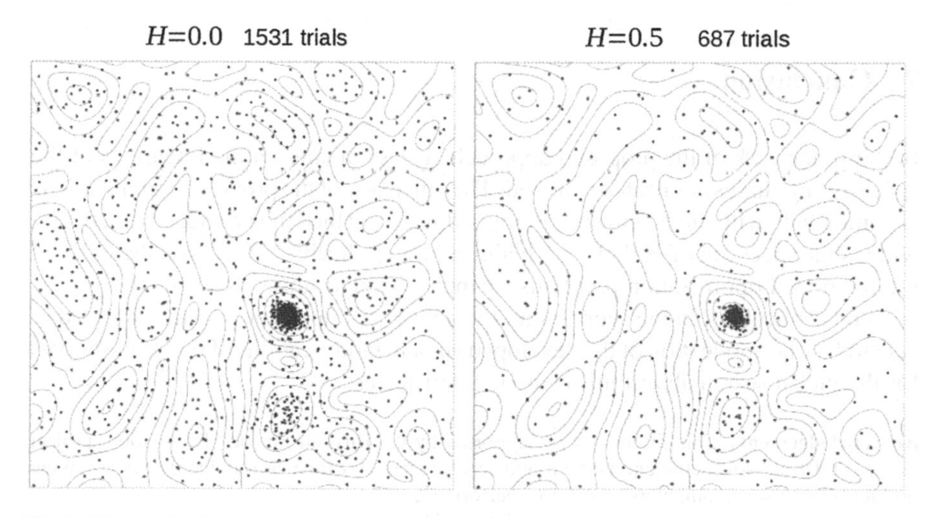

Fig. 5 The results of solving one of the problems using the Peano curve and the smooth modification

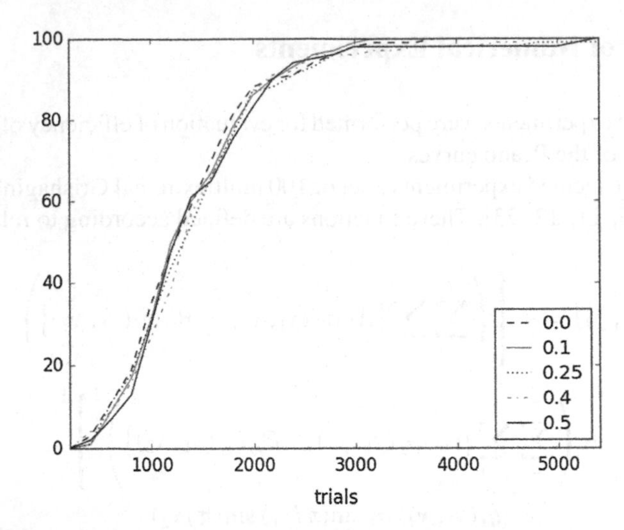

Fig. 6 Operating characteristics of the global search algorithm for the Peano curve and its smooth modifications

For more comprehensive comparison of the methods of constructing the Peano curves, 100 optimization problems from the set (15) were solved. According to the results of the experiments, the operational characteristics of GMAGS were built [5, 11, 12, 23].The operational characteristic of an optimization method is a curve, which shows the dependency of number of solved problems versus the number of iterations. The operational characteristics of GMAGS with using the Peano curve and its smooth modification are presented in Fig. 6.

7 Conclusion

In the framework of the proposed approach, a class of smooth modifications of the space-filling curves has been developed. The efficiency of the proposed modifications is comparable with the Peano curves applied widely. Thus, the provided smoothness widens the spectrum of global search algorithms applicable for solving the reduced multiextremal optimization problems (namely, the algorithms using the derivatives become applicable). In the scope of all the above, an important direction for further research is the application of the developed smooth modifications of the Peano curves for the global search algorithms using the derivatives.

Acknowledgements This research was supported by the Russian Science Foundation, project No 16-11-10150 Novel efficient methods and software tools for time-consuming decision-making problems using supercomputers of superior performance.

References

1. Trn, A., Ilinskas, A.: Global Optimization. Lecture Notes in Computer Science 350. Springer, Heidelberg (1989)
2. Horst, R., Tuy, H.: Global Optimization: Deterministic Approaches. Springer, Heidelberg (1990)
3. Zhigljavsky, A.A.: Theory of Global Random Search. Kluwer Academic Publishers, Dordrecht (1991)
4. Pintr, J.D.: Global Optimization in Action (Continuous and Lipschitz Optimization: Algorithms, Implementations and Applications). Kluwer Academic Publishers, Dordrecht (1996)
5. Strongin, R.G., Sergeyev, Y.D.: Global Optimization with Non-convex Constraints: Sequential and Parallel Algorithms. Kluwer Academic Publishers, Dordrecht (2000)
6. Locatelli, M., Schoen, F.: Global optimization: theory, algorithms, and applications. SIAM (2013)
7. Floudas, C.A., Pardalos, M.P.: Recent Advances in Global Optimization. Princeton University Press, Princeton (2016)
8. Chendes, T. (ed.).: Development in Reliable Computing. Kluwer Academic Publishers, Dordrecht (1999)
9. Sergeyev, Y.D., Strongin, R.G., Lera, D.: Introduction to Global Optimization Exploiting Space-Filling Curves. Springer, Heidelberg (2013)
10. Paulaviius, R., Ilinskas, A.: Simplicial Global Optimization. Springer Briefs in Optimization. Springer, New York (2014)
11. Strongin, R.G.: Numerical Methods in Multi-extremal Problems (Information-Statistical Algorithms). Nauka, Moscow (1978). (in Russian)
12. Strongin, R.G., Gergel, V.P., Grishagin, V.A., Barkalov, K.A.: Parallel Computations for Global Optimization Problems. Moscow State University, Moscow (2013). (in Russian)
13. Barkalov, K.A., Gergel, V.P.: Multilevel scheme of dimensionality reduction for parallel global search algorithms. In: Proceedings of the 1st International Conference on Engineering and Applied Sciences Optimization, pp. 2111–2124 (2014)
14. Gergel, V., Grishagin, V., Israfilov, R.: Local tuning in nested scheme of global optimization. Procedia Comput. Sci. **51**, 865–874 (2015)
15. Gergel V.P., Grishagin V.A., Gergel A.V.: Adaptive nested optimization scheme for multidimensional global search. J. Glob. Optim., 1–17 (2015)
16. Gergel, V., Sidorov, S. .A: Two-level parallel global search algorithm for solution of computationally intensive multiextremal optimization problems. In: Malyshkin, V. (Ed.) PaCT 2015, LNCS, vol. 9251, pp. 505–515. Springer, Heidelberg (2015)
17. Lera, D., Sergeyev, Y.D.: Acceleration of univariate global optimization algorithms working with lipschitz functions and lipschitz first derivatives. SIAM J. Optim. **23**(1), 508–529 (2013)
18. Sergeyev, Y.D., Kvasov, D.E.: A deterministic global optimization using smooth diagonal auxiliary functions. Commun. Nonlinear Sci. Numer. Simul. **21**(1–3), 99–111 (2015)
19. Kvasov, D.E., Sergeyev, Y.D.: Deterministic approaches for solving practical black-box global optimization problems. Adv. Eng. Softw. **80**, 58–66 (2015)
20. Breiman, L., Cutler, A.: A deterministic algorithm for global optimization. Math. Program. **58**(1–3), 179–199 (1993)
21. Gergel, V.P.: A method of using derivatives in the minimization of multiextremum functions. Comput. Math. Math. Phys. **36**(6), 729–742 (1996)
22. Sergeyev, Y.D.: Global one-dimensional optimization using smooth auxiliary functions. Math. Program. **81**(1), 127–146 (1998)
23. Grishagin, V.A.: Operating characteristics of some global search algorithms. Probl. Stoch. Search. **7**, 198–206 (1978). (In Russian)

Iterative Local Search Heuristic for Truck and Trailer Routing Problem

Ivan S. Grechikhin

Abstract Vehicle Routing Problem is a well-known problem in logistics and transportation. There is a big variety of VRP problems in the literature, as they arise in many real-life situations. It is a NP-hard combinatorial optimization problem and finding an exact optimal solution is practically impossible in real-life formulations. There is an important subclass of VRP, which is called Truck and Trailer Routing Problem. For this class of problems, every vehicle contains truck and, possibly, trailer parts. In this work, Site-Dependent Truck and Trailer Routing Problem with Hard and Soft Time Windows and Split Deliveries are considered. We develop an Iterative Local Search heuristic for solving this problem. The heuristic is based on the local search approach and also allows infeasible solutions. A greedy heuristic is applied to construct an initial solution.

Keywords Truck and trailer routing problem · Site-dependent · Soft time windows · Split deliveries · Local search

1 Introduction

Vehicle Routing Problem is a well-known problem in combinatorial optimization and integer programming [1]. The problem can be described as follows: there is a set of customers, where each customer has a demand, there is a set of vehicles, which serve the demand of customers. The demand is usually understood as the set of goods to deliver. The solution of the problem is a set of routes, where each route is served by one vehicle. Also, every route starts and ends in a distribution depot—the point, where the goods are situated initially. Using the information on travel time and cost of travelling between each pair of customers, the goal is to find the solution with

I.S. Grechikhin (✉)
National Research University Higher School of Economics,
Laboratory of Algorithms and Technologies for Network Analysis,
136 Rodionova St, Nizhny Novgorod 603093, Russia
e-mail: igrechikhin@hse.ru

V.A. Kalyagin et al. (eds.), *Models, Algorithms, and Technologies for Network Analysis*, Springer Proceedings in Mathematics & Statistics 197,
DOI 10.1007/978-3-319-56829-4_6

minimal total cost. This paper considers one version of the problem, which is called Truck and Trailer Routing Problem (TTRP). The considered problem is a real-life problem with a big number of constraints.

Truck and trailer routing problem [3] has two sets of customers: truck-customers and trailer-customers. Every vehicle has truck and a trailer of some capacities, and every truck has only one trailer it may utilize (the trailer capacity can be zero, which means that the vehicle does not have a trailer). Truck-customers cannot be served by a vehicle with a trailer. It means that if the vehicle visits a truck-customer or the number of truck-customers, it should not have a trailer from the start of the route, or the trailer should be left at another place before visiting a truck-customer. This requirement is explained by the fact that there may be small stores, that do not have place for a vehicle with its trailer. A vehicle with a trailer has a possibility to leave the trailer at a transshipment location, which is basically a special place to leave trailers. Another opportunity is to leave trailer at some trailer-customer. In this case the trailer may be unloaded at the trailer-customer and, at the same time, the truck goes to some truck-customers. In some routes, the total weight of goods for truck-customers is more than the capacity of the truck without trailer. It justifies a necessity for load transfer—the operation, where goods are transferred from truck to trailer or vice versa.

In the considered problem, the Heterogeneous Fleet of vehicles is used (HFTTRP). It means that vehicles have different capacities and fixed costs, which makes the problem more difficult. Additionally, every customer may have its own limitations on types of vehicles to serve it. In this case the problem is called the Site-Dependent TTRP (SDTTRP) and there are some heuristics developed for such problems. Various Tabu-search heuristic were suggested for the problem [2–4].

Other real-life constraints are hard and soft time windows and split-deliveries. Every customer in the problem has a hard time window and soft time window. Hard time window determines the period of time, when the delivery to the customer is possible. Soft time window defines the preferable time of the delivery to the customer. Lin et al. [5] suggested simulated annealing heuristic for the problem with time window constraints. The problem with heterogeneous fleet and time windows was described by Drexl [6].

Split-deliveries allow to use more than one vehicle to fulfill the delivery to the customer. The whole problem is described as Site-Dependent Truck and Trailer Routing Problem with Time Windows and Split Deliveries (SDTTRPTWSD). The SDTTRPTWSD is considered in Batsyn and Ponomarenko [7, 8]. These papers suggested a greedy heuristic for the problem.

In this article, another approach for solving SDTTRPTWSD is used. Suggested greedy heuristic is used to obtain an initial solution. After the greedy heuristic, the obtained solution is reconstructed by new iterative local search heuristic.

2 The Algorithm Description

The algorithm can be described as follows. First, greedy heuristic constructs a solution iteratively until there are no unserved customers. For every route the algorithm randomly chooses one of the farthest customers to be the first customer added to the route. Then other customers are tried as candidates to the route. The solution has a constraint on the number of split-deliveries and violations of soft time window. For every route, the possibility of a split-delivery and the number of violations are chosen randomly. A route may have only one new split-delivery, and the probability of performing a split-delivery is determined by the estimated number of split-deliveries for not yet served customers. The number of violations is defined similarly by the estimated number of violations for not yet served customers. However, every route may have different number of delays. The algorithm determines the allowed number of violations before constructing the route. After the solution is obtained the iterative local search heuristic tries to move customers between routes to derive a better solution. The algorithm uses a set of changing parameters, which define "neighbourhood"—the region of solutions that are close to the current solution. The heuristic looks through the infeasible solutions and the degree of infeasibility is determined by the parameters—the number of allowed violations over limit, the number of routes with violated capacity, and maximum allowed cost increase of the solution.

2.1 Parameters of the Algorithm

The following parameters are used in the pseudo-code of the algorithm:
V—the set of all customers
K—the set of all vehicles
K_i—the set of vehicles, which can serve customer i
Q_k—the current remaining capacity of vehicle k
q_i—the current remaining demand of customer i
v_R—the number of soft time window violations in route R
R—the current route
S—the current solution
S^*—the best solution found so far
v—the number of permitted soft time window violations
w—the current remaining number of permitted soft time window violations
U—the set of unserved customers sorted by the most expensive (farthest) customer first
C—the cost of the current insertion
Δ—the allowed level of violations in iterative local search
s—the current state
δ—the radius of the neighborhood considered in the local search
$\tilde{S}$—the set of routes, where the capacity of vehicles is violated
$[x_i]$—an array elements x_i.

2.2 Greedy Heuristic

The idea of the greedy algorithm is based on the paper of Solomon [9].

Algorithm 1 Initial Greedy Heuristic

```
1: function INITIALGREEDYHEURISTIC
2:     ▷ Creates an initial feasible solution
3:     U ← V                      ▷ sorting customers so that U_1 has maximal travel cost
4:     S ← ∅
5:     w = v
6:     while U ≠ ∅ do
7:         if NOTPOSSIBLETOSERVE then
8:             goto 3
9:             ▷ There is a small possibility that at some point unserved customers
10:            ▷ are impossible to serve. In that case, the algorithm starts
11:            ▷ the procedure all over again.
12:        end if
13:        i ←UNIFORMRANDOM(U_1, ..., U_μ)        ▷ choose from the μ most expensive
14:        k ←CHOOSEVEHICLE
15:        R ←BASICROUTE
16:        v_R^max ←FINDNUMBEROFVIOLATIONS
17:        R_* ← ∅,  R'_* ← ∅
18:        C_* ← ∞  C'_* ← ∞
19:        i_* ← 0,  i'_* ← 0
20:        repeat
21:            for j ∈ U do
22:                if k ∉ K_j then
23:                    continue
24:                end if
25:                INSERTCUSTOMER
26:                INSERTCUSTOMER
27:            end for
28:            FINALINSERTION
29:        until R'_* = null  &  R_* = null
30:    end while
31:    return S
32: end function
```

Algorithm 2 Insertion of Customer and comparison with best known

```
1: function INSERTCUSTOMER(mayViolate, j, R, v_R^max, C, R_*, i_*)
2:     C_temp ←GETINSERTIONCOST
3:     if C_temp < C then
4:         C ← C_temp
5:         R_* ←GETROUTEWITHCUSTOMER
6:         i_* ← j
7:     end if
8: end function
```

Algorithm 3 Final Insertion Procedure

```
1: function FINALINSERTION(R_*, R'_*, C_*, C'_*, i_*, i'_*, S, w, U)
2:     if R'_* = null & R_* = null then
3:         S ← S ∪ {R}
4:         Q_k ← 0
5:         w ← w − v_R
6:         break
7:     else if R_* = null || (R'_* ≠ null & C'_* < C_*) then
8:         R ← R'_*
9:         U ← U/{i'_*}
10:    else
11:        R ← R_*
12:        U ← U/{i_*}
13:    end if
14: end function
```

The first function of the algorithm is the initial greedy heuristic, which constructs an initial solution (Algorithms 1). The function constructs a solution, which is guaranteed to be feasible, but its cost may not be satisfactory. First, the algorithm sorts all customers by the distance from the depot, so the first customer in U is the farthest one. Then, the process of solution construction begins. Routes of a solution are constructed in a loop, until there are no unserved customers. For every route the algorithm chooses one of the μ farthest customers with the call to UNIFORMRANDOM function, μ is defined before starting the program and usually equals 5. Then a vehicle is determined for the route using CHOOSEVEHICLE function. After that, BASICROUTE creates the route R with one chosen customer i served by the vehicle k.

The function FINDNUMBEROFVIOLATIONS determines the limit of soft window violations for the current route. The probability of allowing one violation depends on the current number of violations and the current number of unserved customers. After the algorithm found the limit of soft window violations, it tries to insert all other customers to route R. It does the insertion in two ways: allowing violations of soft time windows and forbidding violations. Insertion is done by the INSERTCUSTOMER function (Algorithm 2). If one of insertions leads to a situation, where the number of soft window violations exceeds the allowed number, the route is considered infeasible and the function GETINSERTIONCOST returns infinite cost for the insertion. From the obtained routes, the cheapest one is chosen by the FINALINSERTION procedure (Algorithm 3). Step by step the algorithm inserts customers in the current route until every other insertion violates one of the constraints.

As it was said, in the Algorithm 2 the insertion is done in two ways. Parameter $mayViolate$ indicates whether the algorithm allows violation of soft time windows done by this insertion. If $mayViolate = true$, the insertion happens at every possible place in the route, right after the delivery to the previous customer has ended. By doing this insertion, the algorithm does not verify that the new delivery starts inside of the soft time window of corresponding customer. As a consequence of such insertion, the number of violations in the new route may significantly increase and be higher than allowed by the parameter v_R^{max}. If $mayViolate = false$, the algorithm performs the insertion only if new delivery starts inside of the soft time window of corresponding customer. To sum up, if the parameter is *true*, the insertion is implemented as the cheapest possible, and if the parameter is *false*, the insertion takes into account soft time window of the inserted customer. Routes that are obtained from insertion with *true* value are cheaper than routes received with *false* value, however there are situations, when there are no feasible routes with parameter *true*, but some feasible routes with parameter *false*.

2.3 Iterative Local Search Heuristic

The second important function, which implements Iterative Local Search (ILS) heuristic is shown in Algorithm 4. Its goal is to improve initial solution S by iteratively performing simple moves. The algorithm constantly repeats elementary steps using function HEURISTICSTEP. At each step there might happen a move, or it may not happen if the move is overly expensive or infeasible. The variety of possible moves depends on the Δ and δ parameters. Parameter s,which is the state of the heuristic, tracks the number of successful sequential calls of the function (steps that resulted in a change of solution). It also tracks changes in the current best and other parameters through functions CHANGESTATEFORSTEPSUCCESS andCHANGESSTATEFORCHANGEINBEST. Based on that information, function CHANGEVIOLATIONSLEVEL changes parameter Δ. In the process of the heuristic, the current solution S becomes infeasible. Function SHOULDOBTAINFEASIBLESOLUTION, uses the state of the heuristic to determine, when the algorithm should try to obtain a feasible solution from the current solution. After that, there are recovery procedures, which will be described later. Function STOPPINGCONDITION stops the heuristic if the state of the algorithm indicates that it is unlikely to significantly improve the best solution from the current solution. The process repeats itself until the STOPPINGCRITERIA is met. At the moment the STOPPINGCRITERIA is a constraint on a total computing time.

At every call of the heuristic step (Algorithm 5), first, the customer is chosen randomly from one of the routes of the current solution (function CHOOSERANDOMCUSTOMER). After that, the chosen customer is tried to be inserted to other routes in such a way that the next or previous customer in this route is close to it: the travel time is less than δ parameter. The close places are determined by FINDPLACEFORINSERTION. Variable p contains all such places of insertion in route R. The move is performed by the call to ADDCUSTOMER and then function FINDMOVECOST finds the cost of the move with penalties for infeasibility. After finding the best move by cost, this move may be performed if it does not violate the constraints, which is determined by Δ variable and estimated in ALLOWMOVE function.

Finally, when the heuristic tries to obtain a feasible solution from the current infeasible, the recovery procedure takes place (Algorithm 4, lines 11–14). Generally, the whole solution is likely to be in infeasible region because of performed moves.

Algorithm 4 Iterative Local Search

```
1: function ITERATIVE LOCAL SEARCH(S,Δ,δ,s)
2:     S_best ← ∅
3:     repeat
4:         S ←INITIALGREEDYHEURISTIC
5:         S* ← S
6:         repeat
7:             success ←HEURISTICSTEP
8:             CHANGESTATEFORSTEPSUCCESS
9:             if SHOULDOBTAINFEASIBLESOLUTION then
10:                ▷ Recovery procedures work here
11:                ROUTESOPTIMIZATION
12:                RECOVERCAPACITYVIOLATIONS
13:                FINALIZEROUTESTIMES
14:                RECOVERSOFTWINDOWVIOLATIONS
15:                if COST < COST & ISFEASIBLE then
16:                    S* ← S
17:                    CHANGESSTATEFORCHANGEINBEST
18:                end if
19:            end if
20:            CHANGEVIOLATIONSLEVEL
21:        until STOPPINGCONDITION
22:        S_best ← S*
23:    until STOPPINGCRITERIA
24:    return S_best
25: end function
```

Algorithm 5 Heuristic Step Algorithm

```
1: function HEURISTICSTEP(S,Δ,δ)
2:     R_i, i, costOfDeletion ←CHOOSERANDOMCUSTOMER
3:     ▷ i is deleted customer, the algorithm also needs the cost of deletion of this customer from
   its current route
4:     C* ← ∞
5:     R* ← ∅
6:     for all R ∈ S do
7:         p ←FINDPLACEFORINSERTION
8:         R' ←ADDCUSTOMER
9:         c ←FINDMOVECOST
10:        if c < C* then
11:            C* = c
12:            R* = R
13:        end if
14:    end for
15:    if ALLOWMOVE then
16:        S ← S \ {i}
17:        R ← R ∪ {i}
18:        CHANGECURRENTVIOLATIONS
19:        return true
20:    end if
21:    return false
22: end function
```

Algorithm 6 Recovery Capacity Violations Procedure

```
1: function RECOVERCAPACITYVIOLATIONS(S)
2:    S̃ ←FINDROUTESWITHCAPACITYVIOLATIONS
3:    DECREASECAPACITYVIOLATIONS
4:    S̃ =FINDROUTESWITHCAPACITYVIOLATIONS
5:    while S̃ ≠ ∅ do
6:       R_cap ←CHOOSERANDOMROUTE
7:       i ←CHOOSECUSTOMERTORECOVERCAPACITY
8:       k ←CHOOSEVEHICLE
9:       R_new ←BASICROUTE
10:      S ← S ∪ {R_new}
11:      R_cap ← R_cap \ {i}
12:      if DOESNOTVIOLATECAPACITY then
13:         S̃ ← S̃ \ R_cap
14:      end if
15:      DECREASECAPACITYVIOLATIONS
16:   end while
17: end function
```

Algorithm 7 Decreasing the Number of Capacity Violations Procedure

```
1: function DECREASECAPACITYVIOLATIONS(S,S̃)
2:    for all R ∈ S̃ do
3:       C_* ← ∞, R_* ← ∅, R'_* ← ∅
4:       for all i ∈ R do
5:          for all R' ∉ S̃ do
6:             c, R, R' ←FINDCOSTOFMOVE
7:             if c < C_* then
8:                C_* ← c, R_* ← R, R'_* ← R'
9:             end if
10:         end for
11:      end for
12:      if C_* ≠ ∞ then
13:         UPDATEROUTES
14:      end if
15:      if DOESNOTVIOLATECAPACITY then
16:         S̃ ← S̃ \ R_*
17:      end if
18:   end for
19: end function
```

In that case, the algorithm needs to decrease the number of soft time window violations and recover over-capacitated routes. The recovery procedures start with ROUTESOPTIMIZATION—the function finds such place in every route that after shifting earlier customers in this route to the left and latter customers to the right, the number of soft window violations do not increase or increase by 1. It allows to recover solution more efficiently as it ensures the maximal possible time period for insertions.

After the optimization, the algorithm recovers capacities of routes (Algorithm 6). It means that the total number of goods in every route should be less or equal to the capacity of the vehicle in the route. The routes with capacity violations are determined by FINDROUTESWITHCAPACITYVIOLATIONS. Function DECREASECAPACITYVIOLATIONS (Algorithm 7) tries to decrease the number of capacity violations through the loop. In this loop, the function tries to take customers away from the routes with capacity violations and insert them into routes without capacity violations. At the same time, these insertions should not violate capacity of the routes. However, in many cases this function cannot get rid of all capacity violations in one pass. That is why after the first call to the function, the algorithm checks whether there are routes with violations. If the answer is yes, the algorithm chooses one customer from the route with violation (lines 6–7 of Algorithm 6) and creates new route with that customer (lines 8–9). Then, the algorithm repeats call to DECREASECAPACITYVIOLATIONS until there are no capacity violations.

Next step in the recovery part of the ILS heuristic algorithm is FINALIZEROUTETIMES. The procedure goes through every route and compresses the total time of the route because after ROUTESOPTIMIZATION in some routes there may be gaps left. The last step is recovering soft time window violations, which works similarly to the RECOVERCAPACITYVIOLATIONS: it finds routes with soft time window violations, it takes customer or customers with the violations away from one route and inserts them in other route, where the insertion does not increase the number of violations.

3 Computational Results

Experiments were performed on real data for which the results of greedy heuristic are known. The data is represented by the sets of customers (500 customers) and vehicles (200 vehicles), and sets of demands for 7 different days. Each day there are somewhere about 300 customers with demand and this set is different for every day. Also soft time windows are different for every selected day.

The column BP contains the value of objective function obtained by the greedy heuristic for this day (Batsyn and Ponomarenko [8]). The third column shows the results of ILS heuristic for the selected days. The ILS heuristic worked for 3 hours for every day. All experiments were conducted on Intel Xeon X5675 machine, with base processor frequency 3.06 GHz and 64 GB of memory (Table 1).

Table 1 Computational results

Day	BP	ILS	Improvement (%)
Day 1	1200000	1155000	4
Day 2	1100000	1100000	0
Day 3	1160000	1100000	5
Day 4	1200000	1140000	5
Day 5	1245000	1220000	2
Day 6	1235000	1225000	1
Day 7	1275000	1175000	8

4 Conclusion

In this paper, a new heuristic has been developed for the Site-Dependent Truck and Trailer Routing Problem with Time Windows and Split Deliveries. The heuristic uses a greedy approach for an initial solution construction and then employs Iterative Local Search heuristic to improve the initial solution. The obtained results are promising as they show improvement in most cases.

The future work will be directed to improve the speed of the algorithm and to get better results as well. One of the ways to improve the algorithm is to use a new neighborhood—swap neighborhood, where two customers from different routes are swapped. Also, there are more constraints that can be relaxed, such as hard time windows and split deliveries.

Acknowledgements This work is partially supported by Laboratory of Algorithms and Technologies for Network Analysis, National Research University Higher School of Economics.

References

1. Toth, P., Vigo, D.: The Vehicle Routing Problem. Society for Industrial and Applied Mathematics, Philadelphia (2002)
2. Semet, F., Taillard, E.: Solving real-life vehicle routing problems efficiently using tabu search. Ann. Oper. Res. **41**(4), 469–488 (1993)
3. Chao, I.: A tabu search method for the truck and trailer routing problem. Comput. Oper. Res. **29**(1), 33–51 (2002)
4. Scheuerer, S.: A tabu search heuristic for the truck and trailer routing problem. Comput. Oper. Res. **33**(4), 894–909 (2006)
5. Lin, S., Yu, V.F., Lu, C.: A simulated annealing heuristic for the truck and trailer routing problem with time windows. Expert Syst. Appl. **38**(12), 15244–15252 (2011)
6. Drexl, M.: Branch-and-price and heuristic column generation for the generalized truck-and-trailer routing problem. J. Quant. Methods Econ. Bus. Adm. **12**(1), 5–38 (2011)
7. Batsyn, M., Ponomarenko, A.: Heuristic for a real-life truck and trailer routing problem. Procedia Comput. Sci. **31**, 778–792 (2014)
8. Batsyn, M., Ponomarenko, A.: Heuristic for site-dependent truck and trailer routing problem with soft and hard time windows and split deliveries. In: Machine Learning, Optimization, and Big Data, Lecture Notes in Computer Science, pp. 65–79 (2015)
9. Solomon, M.M.: Algorithms for the vehicle routing and scheduling problem with time window constraints. Oper. Res. **35**, 254–265 (1987)

Part II
Network Models

Power in Network Structures

Fuad Aleskerov, Natalia Meshcheryakova and Sergey Shvydun

Abstract We consider an application of power indices, which take into account preferences of agents for coalition formation proposed for an analysis of power distribution in elected bodies to reveal most powerful (central) nodes in networks. These indices take into account the parameters of the nodes in networks, a possibility of group influence from the subset of nodes to single nodes, and intensity of short and long interactions among the nodes.

1 Introduction

The interest in analysis of power in networks has been arisen dramatically for the last years. There have been developed many indices to measure the centrality level of each node. Measures based on the number of links to other nodes are common ones [1]. Other techniques consider how close each node is located to other nodes of the network in terms of the distance, or how many times it is on the shortest paths connecting any given node-pairs [2–5]. Some measures additionally take into account the number of links of adjacent nodes [6–8]. There are also some indices based on cooperative game theory [9]. These indices are called centrality measures.

Unfortunately, most classical measures do not take into account individual properties of each element. Additionally, they do not completely take into account the intensities of interactions between elements, especially, long-range interactions. One

F. Aleskerov (✉) · N. Meshcheryakova · S. Shvydun
National Research University Higher School of Economics (HSE), Moscow, Russia
e-mail: alesk@hse.ru

N. Meshcheryakova
e-mail: natamesc@gmail.com

S. Shvydun
e-mail: shvydun@hse.ru

F. Aleskerov · N. Meshcheryakova · S. Shvydun
V.A. Trapeznikov Institute of Control Sciences of Russian Academy of Sciences (ICS RAS), Moscow, Russia

V.A. Kalyagin et al. (eds.), *Models, Algorithms, and Technologies for Network Analysis*, Springer Proceedings in Mathematics & Statistics 197,
DOI 10.1007/978-3-319-56829-4_7

more problem arises from the fact that not only one node but also a group of nodes can influence other nodes. Consequently, the results of the application of classical measures inadequately represent the actual state of a system.

In [10] a novel method proposed for estimating the intensities of nodes interactions. The method is based on the power index analysis from voting theory that was developed in [11]. A distinct feature of the index is that it takes into account the possibility of the group influence to each individual node. Additionally, it considers only influential edges in the network. The choice of edges that are influential in a network depends on additional parameter q_i which varies with the node i and represents some critical threshold value. However, the Short-Range Interaction Centrality (SRIC) index (originally called a Key Borrower Index) also has some shortages. In the SRIC index only direct interactions of the first level are taken into account, which is not correct in some cases when long-range interactions play a pivotal role or where chain reactions are possible. Also, the SRIC index does not elucidate nodes that have a weak direct influence to particular node i but are highly influential to its adjacent nodes.

The main objective of our research is to improve the index proposed in [10] and develop new efficient methods of key nodes detection. The main feature of our method is that it allows to consider long-range interactions between nodes where the range of influence is a parameter that depends on the problem. Various applications of our method have been already performed for the network of countries those of religion, migration, foreign claims, conflicts, exchange of students, food export–import, etc. It allows to detect hidden influential elements in the networks.

The paper is organized as follows. First, we formally provide the problem statement. Second, we propose new methods for estimation of power in network structures. Finally, we give a simple example and conclude.

2 Power in Network Structures by Long-Range Interactions

2.1 Problem Statement

Consider network-graph $G = \{V,E,W\}$, where $V = \{1,n\}$ is the set of nodes, $|V| = N$, $E \subseteq V \times V$ is the set of edges, and $W = \{w_{ij}\}$ is the set of weights—real numbers prescribed to each edge $(i, j) \in E$. Network-graph G is directed if $\forall i, j \in V : (i, j) \in E \Rightarrow (j, i) \notin E$. The graph is called unweighted if every edge has the same weight. Below we consider only directed weighted graphs, i.e., the set of pairs $(i, j) \in E$ is ordered.

A network-graph G can also be represented in the form of matrix $W = [w_{ij}]_{NxN}$, where w_{ij} is a weight that indicates the intensity of connection of node i to node j. The matrix W is called a weighted adjacency matrix of the network-graph G. In terms of influence, if $w_{ij} > 0$ then node i influences node j with power w_{ij}, otherwise, node i does not influence node j ($w_{ij} = 0$).

Let $\overleftarrow{N}_i$ be a set of directly connected nodes of node i (incoming neighbors), i.e., $\overleftarrow{N}_i = \{j \in V | w_{ij} \neq 0\}$. Let every node has an individual attribute—predefined threshold q_i, i.e., the threshold level when a node becomes affected. Assume that a group of neighbors of node i $O(i) \subseteq \overleftarrow{N}_i$ is critical if its total influence is more than some predefined threshold q_i and node $k \in O(i)$ is pivotal if his exclusion makes the group noncritical. Then $O_p(i)$ is a set of pivotal nodes in group $O(i)$.

The problem lies in estimation of influence of each node in the network.

2.2 *Long-Range Interaction Centrality (LRIC) Index*

Construct intensity matrix $C = [c_{ij}]$ with respect to weights w_{ij}, thresholds q_i and critical groups $O(j)$ as

$$c_{ij} = \begin{cases} \dfrac{w_{ij}}{\min\limits_{\Omega(j) \subseteq N_j | i \in \Omega_p(j)} \sum_{l \in \Omega(j)} w_{lj}}, & if\ i \in \Omega_p(j) \subseteq \overleftarrow{N}_j, \\ 0, & i \notin \Omega_p(j) \subseteq \overleftarrow{N}_j. \end{cases}$$

The interpretation of the matrix C is the following. If $w_{ij} \geq q_j$ then the direct influence of node i on node j is maximal and is equal to 1. Conversely, if node i does not have a direct connection to node j or it does not belong to any critical group then its direct influence is equal to 0. In other cases, if $0 < w_{ij} < q_j$ but if node i is pivotal for node j then its direct influence is the maximal possible direct influence of node i on node j and equal to c_{ij}, $0 < c_{ij} < 1$.

Thus, we can estimate the direct influence between any pair of nodes of the network. To evaluate indirect influences of nodes, we need to consider all possible paths between a pair of nodes. Let us remind that a path between nodes i and j is a sequence of edges $i, j_1, \ldots, j_k, j$ such that $ipj_1, j_1pj_2, \ldots, j_{k-1}pj_k, j_kpj$, where $j_1pj_2 \Leftrightarrow c_{j1j2} > 0$.

Consider now all possible paths without cycles. Denote by $P^{ij} = \{P_1^{ij}, \ldots, P_m^{ij}\}$ a set of all simple paths between i and j, where m is the total number of simple paths, and $n(k) = |P_k^{ij}| \leq s$ is the length of k-th path. Then the influence of i on j via k-th path P_k^{ij} is defined as

$$f(P_k^{ij}) = c_{il_1^k} \times c_{l_1^k l_2^k} \times \ldots \times c_{l_{n(k)-1}^k j} \tag{1}$$

or

$$f(P_k^{ij}) = \min(c_{il_1^k}, c_{l_1^k l_2^k}, \ldots, c_{l_{n(k)-1}^k j}), \tag{2}$$

where $i, l_1^k, l_2^k, \ldots, l_{n(k)-1}^k, j$ is an ordered sequence of nodes in the k-th path.

Formulae (1) and (2) have the following interpretation. According to the formula (1) the influence of node i on node j through the k-th path P_k^{ij} is calculated as the aggregate value of direct influences between nodes which lie in this path. The formula (2) can be interpreted as the k-th path capacity of the influence.

Since there may be many paths between a pair of nodes, the problem of their aggregation should be considered. We propose two ways of the aggregation of the possible influence; the aggregated results form new matrix $C^* = [c_{ij}^*]$:

1. *The total influence via the sum of possible influences*

$$c_{ij}^*(s) = \min\left\{\sum\nolimits_{k:\left|P_k^{ij}\right|\le s} f(P_k^{ij}),\ 1\right\} \tag{3}$$

2. *The total influence via maximum possible influence*

$$c_{ij}(s) = \max\nolimits_{k:|P_k^{ij}|\le s} f(P_k^{ij}). \tag{4}$$

As a result, we propose two methods of path power estimation and two methods of the aggregation of possible influence. Hence, we can assume that there are four ways of the estimation of long-range interactions in a graph. However, not all combinations of formulae (1)–(2) and (3)–(4) are reasonable. Thus, we propose the following combinations (see Table 1)

- (1)–(3): an influence of i on j goes through all paths with respect to all layers in these paths;
- (1)–(4): an influence of i on j goes through a maximal path with respect to all layers in this path;
- (2)–(4): an influence of i on j goes through a maximal path with respect to one minimal layer in this path;

The combination (2)–(3) is not very reasonable because, first, for every path we evaluate the capacity of the influences which can confine on one edge for different paths; then we sum up these influences, which means that we may consider the same influence several times.

After the total intensity of connection between node i and its adjacent nodes is calculated, the index is aggregated over all nodes taking into account the individual attributes of each node.

Table 1 Possible combinations of methods for indirect influence

Path influence\Paths aggregation	Sum of paths influences	Maximal path influence
Multiplication of direct influence	SumPaths	MaxPath
Minimal direct influence	–	MaxMin

3 An Example

Consider a complex system of interconnections from Fig. 1. Assume that threshold level q_i is 25% for each node i, i.e., node i is influenced by individual node or a group of them only if their total influence to i is more than or is equal to 25% of the total influence to i.

Let us calculate some existing centrality measures as well as LRIC index. The results are provided in Table 2.

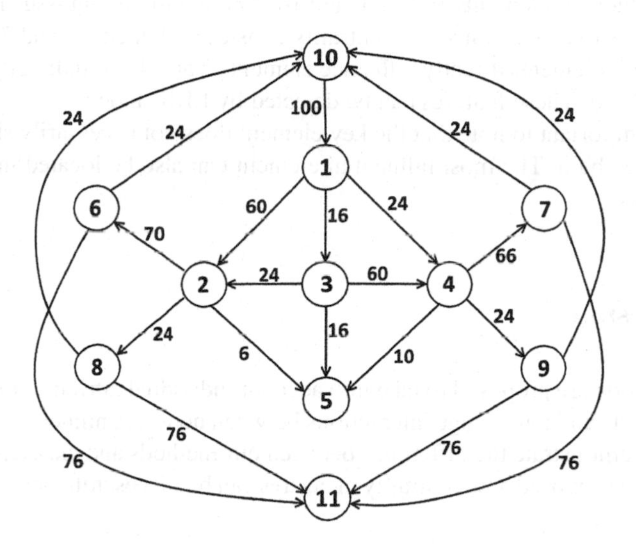

Fig. 1 Numerical example 1

Table 2 Centrality measures for numerical example 1

ID	Degree centrality	Closeness centrality	Betweenness centrality	PageRank	Eigenvector centrality	SRIC	LRIC
1	**200**	**0.0014**	**45**	0.61	**0.11**	**0.11**	0.03
2	184	**0.0012**	**23**	0.57	**0.1**	0.1	0.08
3	116	0.001	0	0.28	0.05	0.01	0.02
4	184	0.001	17	0.47	0.08	**0.11**	0.09
5	32	0.001	0	0.07	0.05	0.01	0.08
6	170	0.001	10	**0.7**	0.095	0.1	**0.11**
7	166	0.001	10	**0.65**	0.08	0.1	**0.12**
8	124	0.001	0	0.56	0.06	0.01	0.1
9	124	0.001	0	0.55	0.05	0.01	0.09
10	**196**	**0.0012**	**43**	0.64	0.09	0	0
11	**304**	**0.002**	0	**1**	**0.22**	**0.44**	**0.27**

As it is shown above LRIC index detect nodes 6, 7, 11 as the most influential of the network. The results are similar to the results of the PageRank centrality, however, the PageRank centrality also highly evaluates the power of elements 1 and 10 while according to our method their influence is rather small. If we accurately analyze the element 10, we will see that this node does not influence any node since its influence is less than the critical threshold value for each node and this element is not pivotal in any groups of nodes. Thus, the influence of the element 1 is overestimated by all indices while the influence of the element 10 should be equal to zero. On the contrary, the influence of the element 5 is underestimated; however, this element directly and indirectly influences elements 1, 2, 3, 4, and 10. Another important issue is that none of these indices except eigenvector centrality considers elements 6 and 7 as pivotal. However, these elements directly influence elements 2 and 4 and indirectly influence elements 1, 3, 10. These features can be detected by LRIC index.

It is also important to note that the key element does not necessarily should be at the end of the chain. The most influential element can also be located in the center of the network.

4 Conclusion

Several methods are proposed to take into account individual attributes of each node as well as short- and long-range interactions between nodes. A numerical example is provided to demonstrate the difference between our methods and classical centrality measures. The proposed new centrality measures can be successfully applied to many real life processes.

Acknowledgements The paper was prepared within the framework of the Basic Research Program at the National Research University Higher School of Economics (HSE) and supported within the framework of a subsidy by the Russian Academic Excellence Project '5-100.' The work was conducted by the International Laboratory of Decision Choice and Analysis (DeCAn Lab) of the National Research University Higher School of Economics.

References

1. Freeman, L.C.: Centrality in social networks: conceptual clarification. Soc. Netw. **1**, 215–239 (1979)
2. Freeman, L.C.: A set of measures of centrality based upon betweenness. Sociometry **40**, 35–41 (1977)
3. Freeman, L.C., Borgatti, S.P., White, D.R.: Centrality in valued graphs: a measure of betweenness based on network flow. Soc. Netw. **13**, 141–154 (1991)
4. Newman, M.E.J.: A measure of betweenness centrality based on random walks. Soc. Netw. **27**, 39–54 (2005)
5. Rochat, Y.: Closeness Centrality Extended to Unconnected Graphs: The Harmonic Centrality Index. ASNA (2009)

6. Bonacich, P.: Technique for analyzing overlapping memberships. Sociol. Methodol. **4**, 176–185 (1972)
7. Brin, S., Page, L.: The anatomy of a large-scale hypertextual Web search engine. Comput. Netw. **30**, 107–117 (1998)
8. Katz, L.: A new status index derived from sociometric index. Psychometrika, pp. 39–43 (1953)
9. Myerson, R.B.: Graphs and cooperation in games. Math. Oper. Res. **2**, 225–229 (1977)
10. Aleskerov, F., Andrievskaya, I., Permjakova, E.: Key Borrowers Detected by the Intensities of Their Short-Range Interactions, Higher School of Economics Research Paper (2014)
11. Aleskerov, F.T.: Power indices taking into account agents' preferences. In: Simeone, B., Pukelsheim, F. (eds) Mathematics and Democracy, pp. 1–18. Springer, Berlin (2006)

Do Logarithmic Proximity Measures Outperform Plain Ones in Graph Clustering?

Vladimir Ivashkin and Pavel Chebotarev

Abstract We consider a number of graph kernels and proximity measures including commute-time kernel, regularized Laplacian kernel, heat kernel, exponential diffusion kernel (also called "communicability"), etc., and the corresponding distances as applied to clustering nodes in random graphs and several well-known datasets. The model of generating random graphs involves edge probabilities for the pairs of nodes that belong to the same class or different predefined classes of nodes. It turns out that in most cases, logarithmic measures (i.e., measures resulting after taking logarithm of the proximities) perform better while distinguishing underlying classes than the "plain" measures. A comparison in terms of reject curves of interclass and intra-class distances confirms this conclusion. A similar conclusion can be made for several well-known datasets. A possible origin of this effect is that most kernels have a multiplicative nature, while the nature of distances used in cluster algorithms is an additive one (cf. the triangle inequality). The logarithmic transformation is a tool to transform the first nature to the second one. Moreover, some distances corresponding to the logarithmic measures possess a meaningful cutpoint additivity property. In our experiments, the leader is usually the logarithmic Communicability measure. However, we indicate some more complicated cases in which other measures, typically, Communicability and plain Walk, can be the winners.

V. Ivashkin (✉)
Moscow Institute of Physics and Technology, 9 Inststitutskii per.,
Dolgoprudny, Moscow Region 141700, Russia
e-mail: vladimir.ivashkin@phystech.edu

P. Chebotarev (✉)
Institute of Control Sciences of the Russian Academy of Sciences, 65 Profsoyuznaya Str.,
Moscow 117997, Russia
e-mail: pavel4e@gmail.com

P. Chebotarev
The Kotel'nikov Institute of Radio-engineering and Electronics (IRE) of Russian Academy of Sciences, Mokhovaya 11-7, Moscow 125009, Russia

V.A. Kalyagin et al. (eds.), *Models, Algorithms, and Technologies for Network Analysis*, Springer Proceedings in Mathematics & Statistics 197,
DOI 10.1007/978-3-319-56829-4_8

1 Introduction

In this paper, we consider a number of graph kernels and proximity measures and the corresponding distances as applied to clustering nodes in random graphs and several datasets. The measures include the commute-time kernel, the regularized Laplacian kernel, the heat kernel, the exponential diffusion kernel, and some others. The model $G(N, (m)p_{\text{in}}, p_{\text{out}})$ of generating random graphs involves edge probabilities for the pairs of nodes that belong to the same class (p_{in}) or different classes (p_{out}). For a review on graph clustering we refer the reader to [14, 16, 28].

The main result of the present study is that in a number of simple cases, logarithmic measures (i.e., measures resulting after taking logarithm of the proximities) perform better while distinguishing underlying classes than the "plain" measures. A direct comparison, in terms of ROC curves, of interclass and intra-class distances confirms this conclusion. However, there are exceptions to that rule. In most experiments, the leader is the new measure logComm (logarithmic Communicability).

Recall that if a proximity measure satisfies the *triangle inequality for proximities* $p(x, y) + p(x, z) - p(y, z) \le p(x, x)$ for all nodes $x, y, z \in V(G)$, then the function $d(x, y) = p(x, x) + p(y, y) - p(x, y) - p(y, x)$ satisfies the ordinary triangle inequality [8]. In this study, we constantly rely on the duality between metrics and proximity measures.

The paper is organized as follows. In the remainder of Sect. 1, we present the metrics and proximity measures under study. In Sect. 2, the logarithmic and plain measures are juxtaposed on several clustering tasks with random graphs generated by the $G(N, (m)p_{\text{in}}, p_{\text{out}})$ model with a small number of classes m. In Sect. 3, 13 measure families compete in two tournaments generated by eight clustering tasks with different parameters. The first tournament gathers the best representatives of each family; the participants of the second one are the representatives with suboptimal parameters corresponding to the 90th percentiles. Every task involves the generation of 50 random graphs. Section 4 presents a different way of comparing the proximity measures: it is based on drawing the ROC curves. This kind of comparison only deals with interclass and intra-class distances and does not depend on the specific clustering algorithm. In Sect. 5, we extend the set of tests: here, the classes of nodes have different sizes, while the intra-class and interclass edge probabilities are not uniform. Finally, in Sect. 6, from random graphs we turn to several classical datasets and make the measure families to meet in two new tournaments. In the concluding Sect. 7, we briefly discuss the results.

Thus, in the following subsections, we list the families of node proximity measures [9], including kernels,[1] and distances, which have been proposed in the literature and have proven to be practical. Generally speaking, our main goal is to find the measures that are the most practical.

[1] On various graph kernels, see, e.g., [15].

1.1 The Shortest Path and Commute-Time Distances

- The **Shortest Path** distance $d^s(i, j)$ on a graph $G = (V, E)$ is the length of a shortest path between i and j in G [1].
- The **Commute Time** distance $d^c(i, j)$ is the average length of random walks from i to j and back. The transition probabilities of the corresponding Markov chain are obtained by normalizing the rows of the adjacency matrix of G. This distance is related to the commute-time kernel [27] $K_{\mathrm{CT}} = L^+$, the pseudoinverse of the Laplacian matrix L of G.
- The **Resistance** distance [19, 23, 29] $d^r(i, j)$ is the effective resistance between i and j in the resistive electrical network corresponding to G.

The resistance distance is well known [2, 18, 25] to be proportional to the commute-time distance.

Let D^s and D^r be the matrices of shortest path distances and resistance distances in G, respectively. As we mainly study parametric families of graph measures, for comparability, the parametric family $(1 - \lambda)D^s + \lambda D^r$ with $\lambda \in [0, 1]$ (i.e., the convex combination of the **Shortest Path** distance and the **Resistance** distance) will be considered. We will denote it by **SP–CT**.

1.2 The Plain Walk, Forest, Communicability, and Heat Kernels/Proximities

Now, we introduce the short names of node proximity measures related to several families of graph kernels.

- **plain Walk** (Von Neumann diffusion kernel, Katz kernel) $K_t^{\mathrm{pWalk}} = (I - tA)^{-1}$, $0 < t < \rho^{-1}$ (ρ is the spectral radius of A, the adjacency matrix of G) [9, 21].
- **Forest** (Regularized Laplacian kernel): $K_t^{\mathrm{For}} = (I + tL)^{-1}$, $t > 0$, where L is the Laplacian matrix of G [7, 10, 30].
- **Communicability** (Exponential diffusion kernel): $K_t^{\mathrm{Comm}} = \exp(tA)$, $t > 0$ [13, 15].
- **Heat kernel** (Laplacian exponential diffusion kernel): $K_t^{\mathrm{Heat}} = \exp(-tL)$, $t > 0$ [11, 24].

1.3 Logarithmic Measures [4]: Walk, Forest, Communicability, and Heat

- **Walk (logarithmic)**: $K_t^{\text{Walk}} = \overrightarrow{\ln K_t^{\text{pWalk}}},\ 0 < t < \rho^{-1}$, where $\overrightarrow{\ln K}$ is the element-wise ln of a matrix K [5].
- **logarithmic Forest**: $K_t^{\text{logFor}} = \overrightarrow{\ln K_t^{\text{For}}},\ t > 0$ [3].
- **logarithmic Communicability**: $K_t^{\text{logComm}} = \overrightarrow{\ln K_t^{\text{Comm}}},\ t > 0.$
- **logarithmic Heat**: $K_t^{\text{logHeat}} = \overrightarrow{\ln K_t^{\text{Heat}}},\ t > 0.$

1.4 Sigmoid Commute Time and Sigmoid Corrected Commute-Time Kernels [16, 31, 32]

The corrected commute-time kernel is defined by

$$K_{\text{CCT}} = HD^{-\frac{1}{2}}M(I-M)^{-1}MD^{-\frac{1}{2}}H \quad \text{with} \quad M = D^{-\frac{1}{2}}\Big(A - \frac{\mathbf{d}\mathbf{d}^T}{\text{vol}(G)}\Big)D^{-\frac{1}{2}},$$

where $H = I - \mathbf{e}\mathbf{e}^T/N$ is the centering matrix, $\mathbf{e} = (\underbrace{1,\ldots,1}_{N})^T,\ \mathbf{d} = A\mathbf{e},\ D = \text{diag}(\mathbf{d}),\ \text{diag}(\mathbf{v})$ is the diagonal matrix with vector $\mathbf{v}$ on the diagonal, and $\text{vol}(G) = |V|,\ V$ being the edge set of G.

Applying the element-wise sigmoid transformation to K_{CT} and K_{CCT} we obtain the corresponding *sigmoid kernels* K^S:

$$[K^S]_{ij} = \frac{1}{1+\exp(-tk_{ij}/\sigma)}, \quad i, j = 1,\ldots,N,$$

where k_{ij} is an element of a kernel matrix (K_{CT} or K_{CCT}), t is a parameter, and σ is the standard deviation of the elements of the kernel matrix. The **Sigmoid Commute-Time kernel** and **Sigmoid Corrected Commute-Time kernel** will be abbreviated as **SCT** and **SCCT**, respectively.

1.5 Randomized Shortest Path and Free Energy Dissimilarity Measures [22]

- **Preliminaries**:

$$P^{\text{ref}} = D^{-1}A, \quad \text{where } D = \text{diag}(A\mathbf{e});$$

$$W = P^{\text{ref}} \circ \overrightarrow{\exp(-tC)}, \quad \text{where } \circ \text{ is element-wise product,}$$

C is the matrix of the shortest path distances, t being the "inverse temperature" parameter;

$$Z = (I - W)^{-1}.$$

- **Randomized Shortest Path (RSP)**:

$$S = (Z(C \circ W)Z) \div Z, \quad \text{where } \div \text{ is element-wise division;}$$

$$\bar{C} = S - \mathbf{e}(\mathbf{d}_S)^T; \quad \mathbf{d}_S = \text{diag}(S), \text{ where diag}(S) \text{ is the vector on the diagonal of square matrix } S;$$

$$\Delta_{\text{RSP}} = (\bar{C} + \bar{C}^T)/2.$$

- **Helmholtz Free Energy distance (FE)**:

$$Z^h = ZD_h^{-1}, \quad \text{where } D_h = \text{Diag}(Z),$$

where $\text{Diag}(Z)$ denotes the diagonal matrix whose diagonal coincides with that of Z;

$$\Phi = -t^{-1} \overrightarrow{\ln Z^h s};$$

$$\Delta_{\text{FE}} = (\Phi + \Phi^T)/2.$$

As we know from the classical scaling theory, the inner product matrix (which is a kernel) can be obtained from a [Euclidean] distance matrix Δ by the

$$K = -\frac{1}{2} H \Delta^{(2)} H$$

transformation, where $H = I - \mathbf{e}\mathbf{e}^T/N$ is the centering matrix.

For comparability, all family parameters are adjusted to the [0, 1] segment by a linear transformation or some $t/(t+c)$ transformation or both of them.

The comparative behavior of graph kernels in clustering tasks has been studied in several papers, including [12, 22, 31]. The originality of our approach is that (1) we do not fix the family parameters and rather optimize them during the experiments, (2) we compare a larger set of measure families, and (3) we juxtapose logarithmic and plain measures.

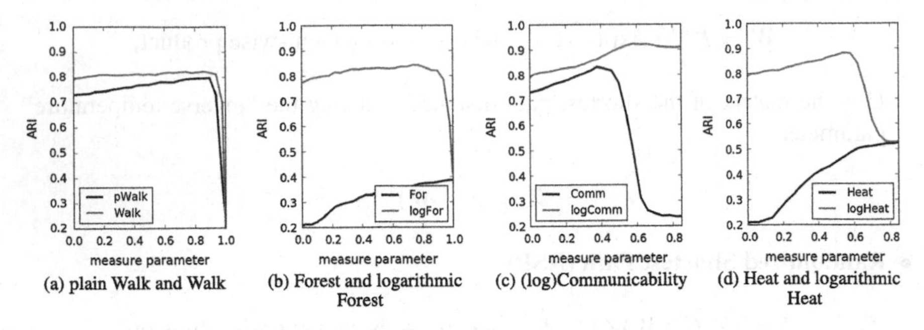

(a) plain Walk and Walk (b) Forest and logarithmic Forest (c) (log)Communicability (d) Heat and logarithmic Heat

Fig. 1 Logarithmic versus plain measures for $G(100, (2)0.2, 0.05)$

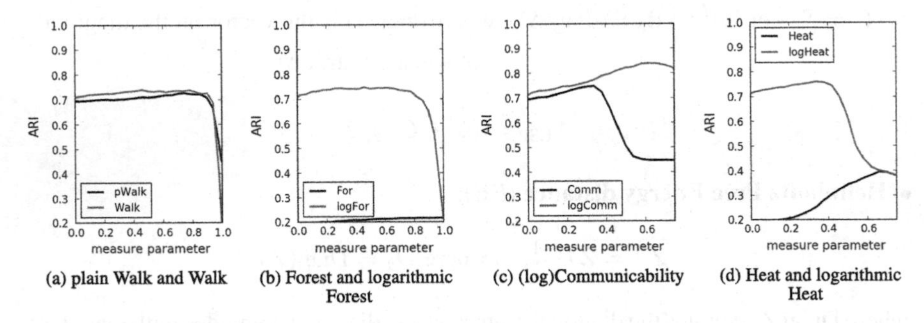

(a) plain Walk and Walk (b) Forest and logarithmic Forest (c) (log)Communicability (d) Heat and logarithmic Heat

Fig. 2 Logarithmic versus plain measures for $G(100, (3)0.3, 0.1)$

2 Logarithmic Versus Plain Measures

Let $G(N, (m)p_{\text{in}}, p_{\text{out}})$ be the model of generating random graphs on N nodes divided into m classes of the same size, with p_{in} and p_{out} being the probability of $(i, j) \in E(G)$ for i and j belonging to the same class and different classes, respectively, where $E(G)$ is the edge set of G.

The curves in Figs. 1, 2 and 3 present the adjusted Rand index[2] (averaged over 200 random graphs) for clustering with Ward's method [33].

It can be seen that in almost all cases, logarithmic measures outperform the ordinary ones. The only exception, where the situation is ambiguous, is the case of Walk measures for random graphs on 200 nodes.

3 Competition by Copeland's Score

In this section, we present the results of many clustering tests in the form of tournaments whose participants are the measure families. Every family is characterized by

[2]On *Rand index* (RI) and *adjusted Rand index* (ARI) we refer to [20].

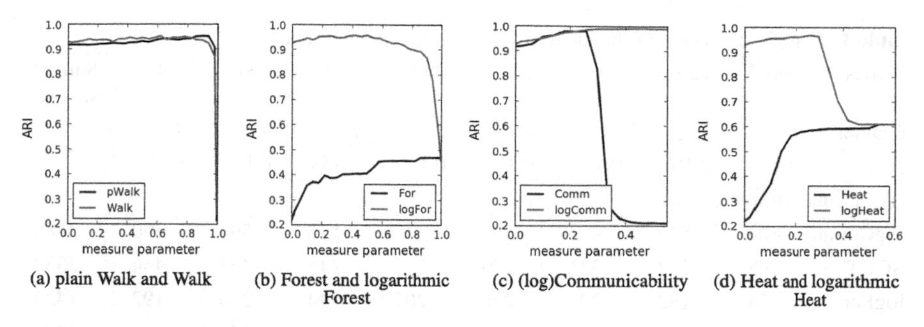

(a) plain Walk and Walk (b) Forest and logarithmic Forest (c) (log)Communicability (d) Heat and logarithmic Heat

Fig. 3 Logarithmic versus plain measures for $G(200, (2)0.3, 0.1)$

its Copeland's score, i.e., the difference between the numbers of "wins" and "losses" in paired confrontations with the other families.

3.1 Approach [22]

- The competition of measure families is based on the paired comparisons.
- Every time when the best adjusted Rand index (ARI) of a measure family F_1 is higher on a random test graph than that of some other measure family F_2, we add $+1$ to the score of F_1 and -1 to the score of F_2.

3.2 The Competition Results

The competition has been performed on random graphs generated with the $G(N, (m)p_{\text{in}}, p_{\text{out}})$ model and the following parameters: $N \in \{100, 200\}$, the number of classes $m \in \{2, 4\}$, $p_{\text{in}} = 0.3$, $p_{\text{out}} \in \{0.1, 0.15\}$. For every combination of parameters, we generated 50 graphs and for each of them we computed the best ARI's the measure families reached. The results are presented in Table 1(a).

3.3 A Competition for 90th Percentiles

Whenever we are looking for the best parameter of a measure family, we compute ARI on a grid of that parameter. In the above competition, we only compared the highest ARI values. Now consider the set of ARI values some measure family provides as a sample and find its 90th percentile. These percentiles become the participants in another tournament. The motivation behind this approach is to take into account the robustness of each family.

Table 1 Copeland's scores of the measure families on random graphs

Nodes	100	100	100	100	200	200	200	200	Sum of scores
Classes	2	2	4	4	2	2	4	4	
p_{out}	0.1	0.15	0.1	0.15	0.1	0.15	0.1	0.15	
(a) Optimal parameters									
logComm	404	539	453	391	235	578	598	590	**3788**
SCCT	298	299	341	275	297	415	454	454	**2833**
logFor	154	182	202	226	207	44	226	192	**1433**
logHeat	249	261	140	28	175	302	251	−64	**1342**
FE	71	88	161	208	77	63	82	160	**910**
Comm	120	9	27	−2	267	138	156	84	**799**
Walk	−42	130	185	126	−44	−42	49	138	**500**
pWalk	−91	−54	−1	64	109	−90	−23	76	**−10**
SCT	−41	−16	−36	−47	−43	−69	−133	−2	**−387**
RSP	−139	−148	−122	17	−67	−166	−194	−162	**−981**
Heat	−31	−339	−515	−513	−148	−123	−458	−505	**−2632**
SP-CT	−399	−365	−250	−186	−469	−450	−414	−366	**−2899**
For	−553	−586	−585	−587	−596	−600	−594	−595	**−4696**
(b) 90th percentiles									
logComm	471	563	497	472	440	588	591	590	**4212**
SCCT	412	448	446	382	470	450	495	498	**3601**
logFor	171	242	275	166	88	185	296	210	**1633**
Walk	−4	229	291	268	48	145	226	320	**1523**
pWalk	19	45	221	232	113	111	188	217	**1146**
FE	−94	91	95	240	−47	56	18	152	**511**
logHeat	342	91	−22	−243	269	156	61	−376	**278**
SCT	−21	−2	−174	−196	56	50	−46	54	**−279**
Comm	−40	−191	2	−58	10	−213	−70	−153	**−713**
SP-CT	−343	−238	−203	−103	−411	−380	−348	−190	**−2216**
RSP	−473	−335	−328	−60	−426	−365	−366	−222	**−2575**
Heat	15	−396	−507	−504	−64	−198	−445	−500	**−2599**
For	−455	−547	−593	−596	−546	−585	−600	−600	**−4522**

The results of the competition for 90th percentiles are given in Table 1(b).

One can notice a number of differences between the orders of families provided by the first competition and the second one. However, in both cases, logarithmic measures outperform the corresponding plain ones. In particular, it can be observed that FE is also a kind of logarithmic measure, as distinct from RSP.

Here, the undisputed leader is logComm. Second place goes to SCCT, a measure which is not logarithmic, but involves even a more smoothing sigmoid transformation.

4 Reject Curves

In this section, we compare the performance of distances (corresponding to the proximity measures or defined independently) in clustering tasks using reject curves.

4.1 Definition

The ROC curve (also referred to as the reject curve) for this type of data can be defined as follows.

- Let us order the distances $d(x, y),\ x, y \in V(G)$ from the minimum to the maximum, where the distance $d(\cdot, \cdot)$ corresponding to a kernel $p(\cdot, \cdot)$ is produced by the $d(x, y) = p(x, x) + p(y, y) - p(x, y) - p(y, x)$ transformation.[3]
- To each $d(x, y)$ we assign a point in the $[0, 1] \times [0, 1]$ square. Its X-coordinate is the share of interclass distances that are less than or equal to $d(x, y)$, the Y-coordinate being the share of intra-class distances (between different nodes) that are less than or equal to $d(x, y)$.
- The polygonal line connecting the consecutive points is the *reject curve* corresponding to the graph.

A better measure is characterized by a reject curve that goes higher or, at least, has a larger area under the curve.

4.2 Results

The optimal values of the family parameters (adjusted to the $[0,\ 1]$ segment) w.r.t. the ARI in clustering based on the Ward's method for three $G(N, (m)p_{\text{in}}, p_{\text{out}})$ models are presented in Table 2.

The optimum chosen on the grid of 50 parameter values is shown as the first number in each of three columns. The second number is the ARI corresponding to the optimum averaged over 200 random graphs. The maximum averaged ARI's are underlined. All of them belong to logComm.

The reject curves for $G(100, (2)0.3, 0.1)$, and the optimal values of the family parameters (w.r.t. the ARI of Ward's method clustering) are shown in Fig. 4. Each subfigure contains 200 lines corresponding to 200 random graphs.

The ε-like bend of several curves (pWalk, Walk, logFor, SCT, RSP, FE) appears because the corresponding measures strongly correlate with the shortest path (SP) distance between nodes. In these experiments, the SP distance takes only a few small values.

---

[3]Recall that a number of distances that correspond to logarithmic measures possess a meaningful cutpoint additivity property [6].

Table 2 Optimal family parameters and the corresponding ARI's

Measure (kernel)	$G(100, (2)0.3, 0.05)$ Opt. parameter, ARI	$G(100, (2)0.3, 0.1)$ Opt. parameter, ARI	$G(100, (2)0.3, 0.15)$ Opt. parameter, ARI
pWalk	0.86, 0.9653	0.90, 0.8308	0.66, 0.5298
Walk	0.86, 0.9664	0.74, 0.8442	0.64, 0.5357
For	1.00, 0.5816	0.98, 0.3671	0.00, 0.2007
logFor	0.62, 0.9704	0.56, 0.8542	0.52, 0.5541
Comm	0.38, 0.9761	0.32, 0.8708	0.26, 0.5661
logComm	0.68, 0.9838	0.54, 0.9466	0.62, 0.7488
Heat	0.86, 0.6128	0.86, 0.5646	0.78, 0.2879
logHeat	0.52, 0.9827	0.40, 0.8911	0.28, 0.5561
SCT	0.74, 0.9651	0.62, 0.8550	0.64, 0.5531
SCCT	0.36, 0.9834	0.26, 0.9130	0.22, 0.6626
RSP	0.99, 0.9712	0.98, 0.8444	0.98, 0.5430
FE	0.94, 0.9697	0.94, 0.8482	0.86, 0.5460
SP-CT	0.28, 0.9172	0.34, 0.6782	0.42, 0.4103

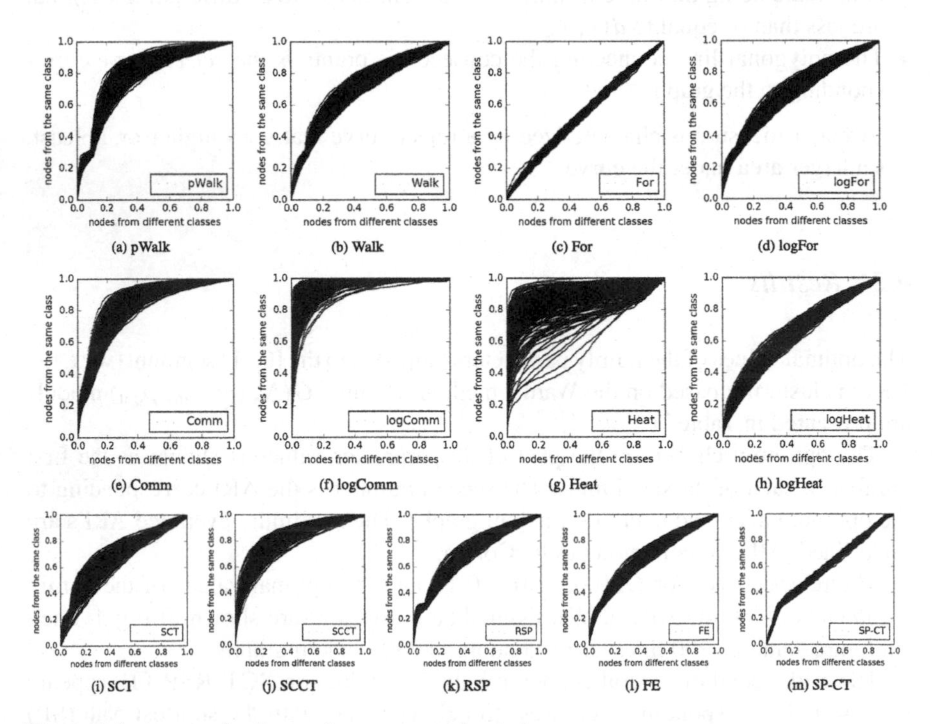

(a) pWalk (b) Walk (c) For (d) logFor

(e) Comm (f) logComm (g) Heat (h) logHeat

(i) SCT (j) SCCT (k) RSP (l) FE (m) SP-CT

Fig. 4 Reject curves for the graph measures under study

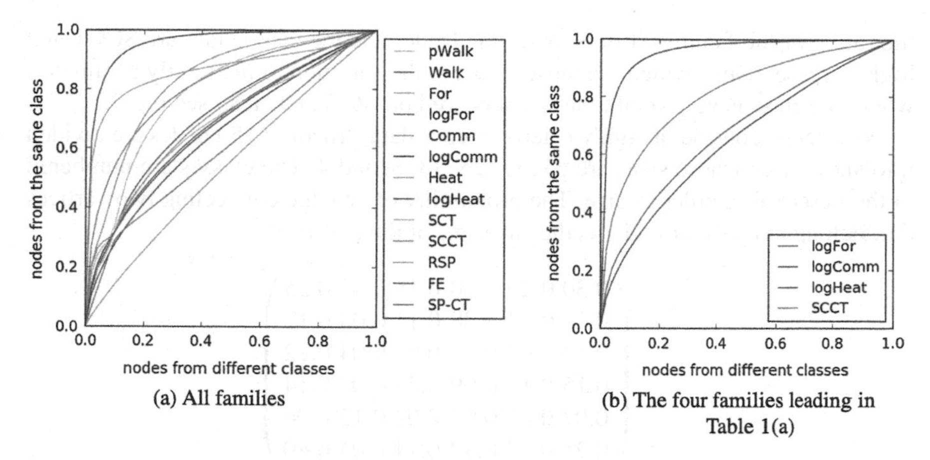

(a) All families

(b) The four families leading in Table 1(a)

Fig. 5 Average reject curves

Finally, in Fig. 5a we show the reject curves averaged over 200 random graphs. The curves for the four families that are leaders in Table 1 are duplicated in Fig. 5b.

One can observe that the results are partially concordant with those obtained with Ward's method. In particular, the first place goes to logComm. Therefore, these results are not an exclusive feature of Ward's method. Notice that Heat has a good average reject curve. However, it produces relatively many large intra-class distances and its partial results are extremely unstable. This supposedly determines the low values of ARI of this measure.

5 Graphs with Classes of Different Sizes

The $G(N, (m)p_{\text{in}}, p_{\text{out}})$ model generates graphs with nodes divided into classes of the same size. We now consider graphs with $N = 100$ nodes divided into two classes of different sizes. The size of the first class, N_1, is shown along the horizontal axis in Fig. 6.

We see that the ARI's of logComm, SCCT, and logHeat have minima at N_1 near 10 or 15. In contrast, the ARI's of Comm and pWalk have larger maxima in the same interval. As a result, the latter two measures outperform the former three (and the other measures under study) at $N_1 \in [8, 19]$. However, if N_1 is very small, then Ward's method with Comm or pWalk seems to engender misrecognition. Thus, this case can be considered as an exception to the rule that "logarithmic measures outperform plain ones": with a moderate size of the smaller class, Comm and pWalk outperform the logarithmic measures (and SCCT in which the sigmoid function is analogous to the logarithmic one as a smoothing transformation).

In all the above experiments, we looked for the optimal values of the family parameters. If the families of measures are used with random parameter values, then the rating of the families differs. Now, the leader and the vice-leader are SCCT and logFor, respectively, which are most robust to the variation of the family parameter; when one class is very small, the winners are For, SCT, and Heat, see Fig. 7.

Now let us consider a highly heterogeneous data structure: 150 nodes are divided into six classes whose sizes are 65, 35, 25, 13, 8, and 4. The classes are numbered in the descending order of size. The probability of an edge connecting two vertices that belong to classes i and j is the entry p_{ij} of the matrix P:

$$P = \begin{pmatrix} 0.30 & 0.20 & 0.10 & 0.15 & 0.07 & 0.25 \\ 0.20 & 0.24 & 0.08 & 0.13 & 0.05 & 0.17 \\ 0.10 & 0.08 & 0.16 & 0.09 & 0.04 & 0.12 \\ 0.15 & 0.13 & 0.09 & 0.20 & 0.02 & 0.14 \\ 0.07 & 0.05 & 0.04 & 0.02 & 0.12 & 0.04 \\ 0.25 & 0.17 & 0.12 & 0.14 & 0.04 & 0.40 \end{pmatrix}.$$

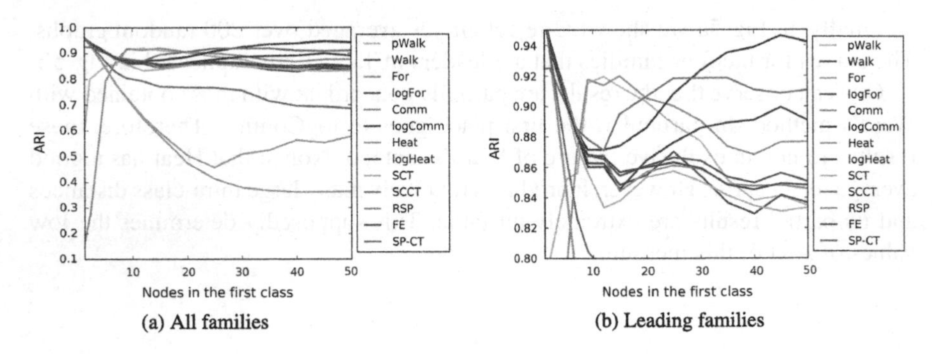

(a) All families (b) Leading families

Fig. 6 Graphs with two classes of different sizes: clustering with optimal parameter values

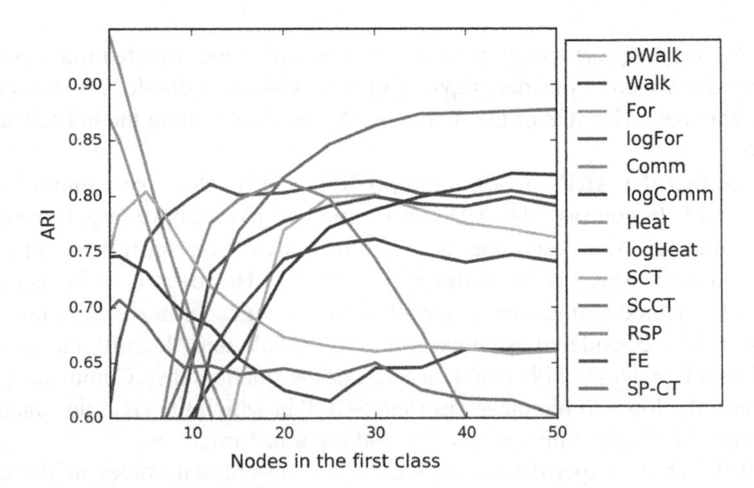

Fig. 7 Graphs with two classes of different sizes: random parameter values

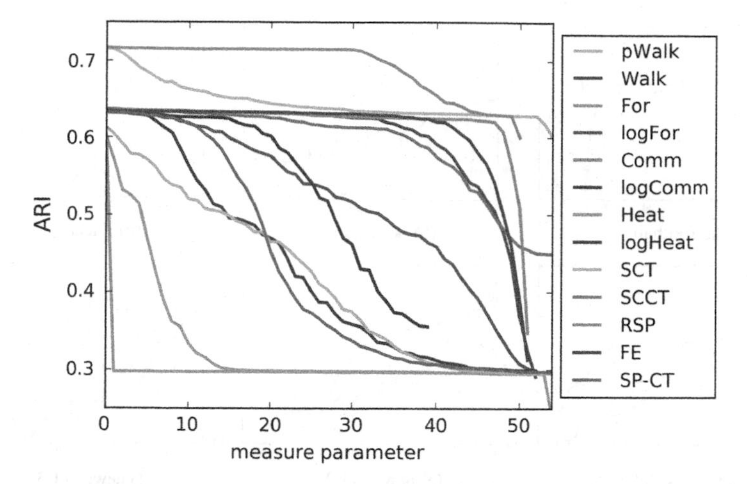

Fig. 8 ARI of various measure families on a structure with six classes

For each measure family, we considered 55 values of the family parameter and sorted them in the descending order of the corresponding ARI averaged for 200 random graphs. ARI against the rank of the family parameter value is shown in Fig. 8. Two things are important for each family: first, the maximum of ARI and second, the velocity of descent.

For this data structure, the leaders are Comm and pWalk, as well as for the two-component graphs with one small, but not very small class of nodes.

6 Cluster Analysis on Several Classical Datasets

Hitherto we mainly considered one type of random graph: the graphs with uniform interclass edge probabilities and uniform intra-class edge probabilities. Certainly, many real-world graphs can hardly be obtained in the framework of that model. In this section, we study clustering on several datasets frequently used to check various graph algorithms.

We investigate a total of nine graphs, the smallest of which (Zachary's Karate club [35]) contains 34 nodes. The largest graph (a Newsgroup graph [34] with three classes) contains 600 nodes. We analyze six Newsgroup datasets. The remaining datasets are Football [17] and Political books [26]. Table 3 presents some features of the datasets.

For each dataset and each measure family, we sorted 55 values of the family parameter in the descending order of the corresponding ARI. ARI against the rank of the family parameter value is shown in Fig. 9.

Finally, we present Copeland's score competition for the measure families: separately for the best values of the family parameters and for 80th percentiles (Tables 4 and 5).

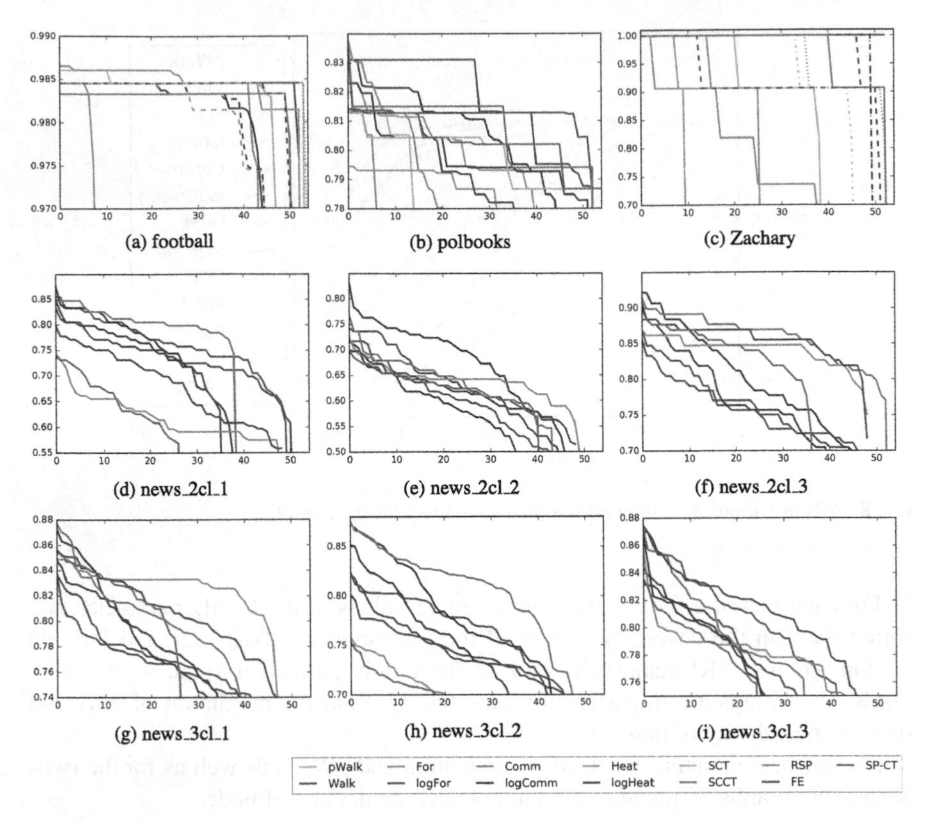

Fig. 9 ARI of various measure families on classical datasets

One can observe that for different datasets, ranking of measure families w.r.t. the quality of clustering differs. In Table 4, for six datasets, SCCT takes the 1st or 2nd place; logComm does so for five datasets. In most cases, the logarithmic measures

Table 3 Overview of the datasets in the experiments

Dataset family	Dataset name	Number of nodes	Number of classes
Football	football	115	12
Political books	polbooks	105	3
Zachary	Zachary	34	2
Newsgroup	news_2cl_1	400	2
	news_2cl_2	400	2
	news_2cl_3	400	2
	news_3cl_1	600	3
	news_3cl_2	600	3
	news_3cl_3	600	3

Table 4 Copeland's scores of the measure families for the best parameter values

	football	polbooks	Zachary	news_2cl_1	news_2cl_2	news_2cl_3	news_3cl_1	news_3cl_2	news_3cl_3	Score
SCCT	−12	12	1	7	10	10	12	12	10	**62**
logComm	−1	5	1	12	12	12	10	10	−2	**59**
logHeat	−1	1	1	7	3	8	2	8	6	**35**
FE	−1	−2	1	2	−1	6	8	0	12	**25**
RSP	−1	10	1	0	6	1	4	−2	4	**23**
Walk	−1	5	1	4	8	−4	0	4	2	**19**
logFor	−1	−6	1	10	3	1	−4	6	8	**18**
SP-CT	−1	8	1	−3	−1	4	6	−4	0	**10**
SCT	−1	−10	1	−3	−4	−2	−2	2	−4	**−23**
Comm	12	−6	1	−6	−6	−6	−6	−8	−8	**−33**
pWalk	10	−6	1	−8	−8	−8	−8	−6	−6	**−39**
Heat	−1	1	1	−10	−10	−11	−11	−10	−11	**−62**
For	−1	−12	−12	−12	−12	−11	−11	−12	−11	**−94**

Table 5 Copeland's scores of the measure families for 80th percentiles

	football	polbooks	Zachary	news_2cl_1	news_2cl_2	news_2cl_3	news_3cl_1	news_3cl_2	news_3cl_3	Score
logComm	0	10	3	10	12	8	4	10	4	**61**
SCCT	−10	8	3	12	8	12	6	12	8	**59**
FE	0	3	3	2	4	8	12	2	12	**46**
Walk	0	12	3	4	10	−4	0	4	6	**35**
logFor	0	3	3	8	4	−2	−2	6	10	**30**
SP-CT	0	3	3	0	−1	8	8	−2	0	**19**
logHeat	0	−7	3	6	−1	2	2	8	2	**15**
RSP	−12	3	−8	−2	4	4	10	0	−2	**−3**
SCT	0	−7	3	−8	−4	0	−4	−4	−4	**−28**
pWalk	11	−4	3	−6	−8	−8	−6	−7	−6	**−31**
Comm	11	−12	3	−4	−6	−6	−8	−7	−8	**−37**
Heat	0	−2	−11	−11	−11	−11	−11	−11	−11	**−79**
For	0	−10	−11	−11	−11	−11	−11	−11	−11	**−87**

outperform the corresponding plain ones. For the "news_3cl_3" dataset and 80th percentiles, the leaders are FE and logFor. For "Zachary" with the best parameter, all measures, except for Forest, reach an absolute result. For "football" (having 12 classes), Comm and pWalk are the winners with the best parameters, like in the cases of two classes of different sizes (cf. Fig. 6b) and of six classes (Fig. 8); SCCT is the worst.

The comparison of Tables 4 and 5 demonstrates that the rather high results of logHeat and RSP are not stable enough, so in the ranking with 80th percentiles, they lose four and three positions, respectively; Walk, logFor and SP-CT shift up two places each. This dynamics resembles that in Table 1(a), (b).

7 Conclusion

The main conclusion of our study is that in most cases, including the simple cases of random graphs with homogeneous classes of similar size, logarithmic measures (i.e., measures resulting after taking logarithm of the proximities) better reveal the underlying structure than the "plain" measures do. A direct comparison of interclass and intra-class distances by drawing the reject curves confirms this conclusion (with the exception of Heat and logHeat).

In our experiments, the three leading measure families in the aforementioned simple cases, according to Copelands's test presented in Table 1, are logarithmic Communicability, Sigmoid Corrected Commute-Time kernel, and logarithmic Forest. The superiority of logarithmic Communicability over the other measures is observed here for all sets of random graphs, except for the set (200, 2, 0.1), for which SCCT is the best.

A plausible explanation of the superiority of logarithmic measures is that most kernels and proximity measures under study have a multiplicative nature, while the nature of distances that cluster algorithms actually use is an additive one (as the triangle inequality reveals). The logarithmic transformation is precisely the tool that transforms the first nature to the second one. Moreover, some distances corresponding to the logarithmic measures possess a meaningful cutpoint additivity property.

At the same time, there are more complex heterogeneous networks for which other measures can behave well. Among such structures, we can mention one type of networks with classes of different sizes and smaller classes of moderate sizes, for which two "plain" measures, Comm and pWalk can outperform the logarithmic measures under study. A similar situation can be observed for some structures with many classes. The SCCT kernel, which involves the sigmoid transformation instead of the logarithmic one, performs very well in many experiments. In Ward's clustering (with the best parameter values) of several datasets, it even wins in the competition by Copeland's score.

Acknowledgements The work of the second author was supported by the Russian Science Foundation [project 16-11-00063 granted to IRE RAS].

References

1. Buckley, F., Harary, F.: Distance in Graphs. Addison-Wesley, Redwood City, CA (1990)
2. Chandra, A.K., Raghavan, P., Ruzzo, W.L., Smolensky, R., Tiwari, P.: The electrical resistance of a graph captures its commute and cover times. In: Proceedings of 21st Annual ACM Symposium on Theory of Computing, pp. 574–586. ACM Press, Seattle (1989)
3. Chebotarev, P.: A class of graph-geodetic distances generalizing the shortest-path and the resistance distances. Discret. Appl. Math. **159**(5), 295–302 (2011)
4. Chebotarev, P.: The graph bottleneck identity. Adv. Appl. Math. **47**(3), 403–413 (2011)
5. Chebotarev, P.: The walk distances in graphs. Discret. Appl. Math. **160**(10–11), 1484–1500 (2012)
6. Chebotarev, P.: Studying new classes of graph metrics. In: Nielsen, F., Barbaresco, F. (eds.) Proceedings of the SEE Conference "Geometric Science of Information" (GSI-2013). Lecture Notes in Computer Science, LNCS, vol. 8085, pp. 207–214. Springer, Berlin (2013)
7. Chebotarev, P.Y., Shamis, E.V.: The matrix-forest theorem and measuring relations in small social groups. Autom. Remote Control **58**(9), 1505–1514 (1997)
8. Chebotarev, P.Y., Shamis, E.V.: On a duality between metrics and Σ-proximities. Autom. Remote Control **59**(4), 608–612 (1998)
9. Chebotarev, P.Y., Shamis, E.V.: On proximity measures for graph vertices. Autom. Remote Control **59**(10), 1443–1459 (1998)
10. Chebotarev, P., Shamis, E.: The forest metrics for graph vertices. Electron. Notes Discret. Math. **11**, 98–107 (2002)
11. Chung, F., Yau, S.T.: Coverings, heat kernels and spanning trees. J. Comb. **6**, 163–184 (1998)
12. Collette, A.: Comparison of some community detection methods for social network analysis. Master's thesis, Louvain School of Management, Universite catholique de Louvain, Louvain, Belgium, 80 p. (2015)
13. Estrada, E.: The communicability distance in graphs. Linear Algebra Appl. **436**(11), 4317–4328 (2012)
14. Fortunato, S.: Community detection in graphs. Phys. Rep. **486**(3), 75–174 (2010)
15. Fouss, F., Francoisse, K., Yen, L., Pirotte, A., Saerens, M.: An experimental investigation of kernels on graphs for collaborative recommendation and semisupervised classification. Neural Netw. **31**, 53–72 (2012)
16. Fouss, F., Saerens, M., Shimbo, M.: Algorithms and Models for Network Data and Link Analysis. Cambridge University Press, Cambridge (2016)
17. Girvan, M., Newman, M.E.J.: Community structure in social and biological networks. Proc. Nat. Acad. Sci. **99**(12), 7821–7826 (2002)
18. Göbel, F., Jagers, A.A.: Random walks on graphs. Stoch. Proces. Appl. **2**(4), 311–336 (1974)
19. Gvishiani, A.D., Gurvich, V.A.: Metric and ultrametric spaces of resistances. Rus. Math. Surv. **42**(6), 235–236 (1987)
20. Hubert, L., Arabie, P.: Comparing partitions. J. Classif. **2**(1), 193–218 (1985)
21. Kandola, J., Cristianini, N., Shawe-Taylor, J.S.: Learning semantic similarity. In: Advances in Neural Information Processing Systems, pp. 657–664 (2002)
22. Kivimäki, I., Shimbo, M., Saerens, M.: Developments in the theory of randomized shortest paths with a comparison of graph node distances. Phys. A Stat. Mech. Appl. **393**, 600–616 (2014)
23. Klein, D.J., Randić, M.: Resistance distance. J. Math. Chem. **12**, 81–95 (1993)
24. Kondor, R.I., Lafferty, J.: Diffusion kernels on graphs and other discrete structures. In: Proceedings of the 19th International Conference on Machine Learning, pp. 315–322 (2002)
25. Nash-Williams, C.S.J.A.: Random walk and electric currents in networks. Math. Proc. Camb. Philos. Soc. **55**(02), 181–194 (1959)
26. Newman, M.E.J.: Modularity and community structure in networks. Proc. Nat. Acad. Sci. **103**(23), 8577–8582 (2006)

27. Saerens, M., Fouss, F., Yen, L., Dupont, P.: The principal components analysis of a graph, and its relationships to spectral clustering. In: Machine Learning: ECML 2004, pp. 371–383. Springer, Cham (2004)
28. Schaeffer, S.E.: Graph clustering. Comput. Sci. Rev. **1**(1), 27–64 (2007)
29. Sharpe, G.E.: Solution of the $(m + 1)$-terminal resistive network problem by means of metric geometry. In: Proceedings of the First Asilomar Conference on Circuits and Systems, pp. 319–328. Pacific Grove, CA (1967)
30. Smola, A.J., Kondor, R.I.: Kernels and regularization of graphs. In: Proceedings of the 16th Annual Conference on Learning Theory, pp. 144–158 (2003)
31. Sommer, F., Fouss, F., Saerens, M.: Comparison of graph node distances on clustering tasks, Lecture Notes in Computer Science, LNCS, vol. 9886, pp. 192–201. Springer, Cham (2016)
32. von Luxburg, U., Radl, A., Hein, M.: Getting lost in space: Large sample analysis of the resistance distance. NIPS 2010. Twenty-Fourth Annual Conference on Neural Information Processing Systems, pp. 1–9. Curran, Red Hook, NY (2011)
33. Ward Jr., J.H.: Hierarchical grouping to optimize an objective function. J. Am. Stat. Assoc. **58**(301), 236–244 (1963)
34. Yen, L., Fouss, F., Decaestecker, C., Francq, P., Saerens, M.: Graph nodes clustering with the sigmoid commute-time kernel: a comparative study. Data Knowl. Eng. **68**, 338–361 (2009)
35. Zachary, W.W.: An information flow model for conflict and fission in small groups. J. Anthropol. Res. 452–473 (1977)

Analysis of Russian Power Transmission Grid Structure: Small World Phenomena Detection

Sergey Makrushin

Abstract In this paper, the complex network theory is used to analyze the spatial and topological structure of the Unified National Electricity Grid (UNEG)—Russia's power transmission grid, the major part of which is managed by Federal Grid Company of the Unified energy system. The research is focused on the applicability of the small-world model to the UNEG network. Small-world networks are vulnerable to cascade failure effects what underline importance of the model in power grids analysis. Although much research has been done on the applicability of the small-world model to national power transmission grids, there is no generally accepted opinion on the subject. In this paper we, for the first time, used the latticization algorithm and small-world criterion based on it for transmission grid analysis. Geo-latticization algorithm has been developed for a more accurate analysis of infrastructure networks with geographically referenced nodes. As the result of applying the new method, a reliable conclusion has been made that the small-world model is applicable to the UNEG. Key nodes and links which determine the small-world structure of the UNEG network have been revealed. The key power transmission lines are critical for the reliability of the UNEG network and must be the focal point in preventing large cascade failures.

1 Introduction

The object of the current research is the Unified National Electricity Grid (UNEG) of Russia, the major part of which is managed by Federal Grid Company of Unified Energy System (FGC UES Inc.). The aims of the research are to find the key network topology properties and to identify the UNEG network model. The main focus is on the detection of the small-world phenomena in the UNEG network and on finding appropriate methods for this analysis. These methods are based on the complex

S. Makrushin (✉)
Financial University Under the Government of the Russian Federation,
49 Leningradsky Prospekt, GSP-3, 125993 Moscow, Russia
e-mail: SVMakrushin@fa.ru

V.A. Kalyagin et al. (eds.), *Models, Algorithms, and Technologies for Network Analysis*, Springer Proceedings in Mathematics & Statistics 197,
DOI 10.1007/978-3-319-56829-4_9

network theory and use for analysis the UNEG computer model. Complex network theory methods are widely employed for power transmission grid analysis all over the world but have not been used for Russian's UNEG.

In the majority of papers on the analysis of power transmission grids with the complex network theory methods the small-world model [23] applicability is considered [13]. This attention can be explained by the fact that the small-world model is one of the crucial models of the complex network theory. One of the major properties of small-world networks is their capability of fast diffusion of information (or any other process) in a network [14]. As concerns electricity transmission networks it means that small-world networks are vulnerable to cascade failure effects what underline importance of the model in power grids analysis [13]. Although much research has been done on the applicability of the small-world model to national power transmission grids, there is no common opinion on that subject [6, 13, 22].

One of the reasons for this is the lack of appropriate methods for the identification of small-world topology in a network. The current paper considers new more accurate methods for the identification problem; these methods rely on the results which were gained by Telesford et al. [21] but have not been used for electrical network analysis. The paper also proposes a modification for that method which is adapted for a network with geographically referenced nodes.

2 The UNEG Network Computer Model

For the complex network analysis UNEG appears as a network with electric power stations and substations as nodes and high-voltage power lines as links. For the current research, data on 514 nodes and 614 links has been gathered. This data includes the networks topology, geographical coordinates of nodes, region binding, voltage levels, and other properties. The data for the computer model creation have been taken from official papers [12], online UNEG map from FGC UES [19], OpenStreetMap GIS [11]. The UNEG computer model has been created for the main operating regions of UNEG except from the unified national energy systems of Siberia, East and Urals (partially). Visualization of the UNEG computer model with nodes in accordance with their geographical coordinates can be seen in Fig. 1a.

The computer model processing and the algorithm development have been done with Python programming language and NetworkX library [10]. Using the computer model of UNEG, the basic properties of the network have been found (see Table 1).

3 The Generally Accepted Small-World Network Criterion

According to the definition, small-world networks have an clustering coefficient close to that of regular lattices with a similar quantity of nodes and links and an average path length close to that of random networks [5] with a similar quantity of nodes

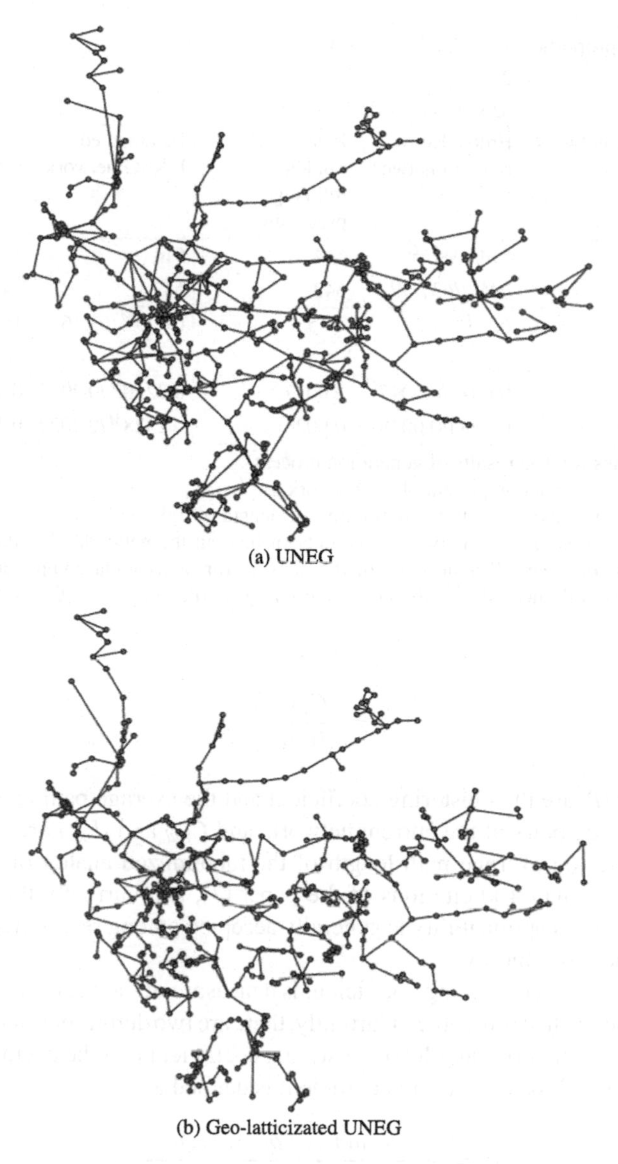

(a) UNEG

(b) Geo-latticizated UNEG

Fig. 1 Geo-latticizated UNEG network **a** Initial UNEG model (nodes shown in accordance with their geographical coordinates) **b** Geo-latticizated UNEG network: if possible, long links are replaced by shorter ones, node's degrees are preserved from changing

and links [23]. In practice, however, to analyze the applicability of the small-world model for a certain network the average path length and the clustering coefficient are compared only between the network and its random analog. There is a widespread small-world model applicability criterion which was presented for the first time in [7]:

Table 1 Basic properties of the UNEG network and its analogs

	1	2	3	4	5
		UNEG network analogs[a]			
	UNEG network	Erdös–Rényi random network	Results of random relinkage procedure	Latticizated UNEG network	Geo-latticizated UNEG network
N	514	514/459.6^G	514	514	514
L	642	645.2/637.5^G	642	642	642
$\langle l \rangle$	11.93	6.51^G	7.35	67.25^M/62.16	16.65^M/17.63
L^{tot}	48107	–	–	–	40237^M/41201
C	0.0942	0.0047/0.0047^G	0.0037	0.2210^M/0.2026	0.1614^M/0.1641
C^{avg}	0.0807	0.0034/0.0038^G	0.0026	0.2328^M/0.2027	0.1771^M/0.1770

[a]The mean values for 100 results of generation process
^{G}The value for the giant component of the network
^{M}The value for the network with the minimum total length of links
N—number of nodes in the network, L—number of links in the network, $\langle l \rangle$—the average path length (in hops) between all node pairs of the network (or of its giant component), L^{tot}—total length of links (in kilometers), C—the global clustering coefficient, C^{avg}—the average clustering coefficient

$$\sigma = \frac{C/C_{rnd}}{\langle l \rangle / \langle l_{rnd} \rangle}, \tag{1}$$

where C and $\langle l \rangle$ are the clustering coefficient and the average path length (in hops) between all node pairs of the current network, and C_{rnd} and $\langle l_{rnd} \rangle$ are the clustering coefficient and the average path length of the randomized analog of the network. There are more than 400 citations of the paper [7], and currently this criterion of small-world model applicability is generally accepted among researchers in the field of complex network theory.

In graph theory, a clustering coefficient is a measure of the degree to which nodes in a graph tend to cluster together. Currently, there are two definitions of the clustering coefficient of a network: the global clustering coefficient and the average clustering coefficient. The global clustering coefficient is defined as:

$$C = \frac{3 \times number\ of\ triangles}{number\ of\ paths\ of\ length\ 2}, \tag{2}$$

where a 'triangle' is a set of three nodes in which each node is connected to the other two. The average clustering coefficient of a network C^{avg} is the mean of local clustering coefficients C_i^{avg} over all nodes, where C_i^{avg} for node i is defined as:

$$C_i^{avg} = \frac{2L_i}{k_i(k_i - 1)}, \tag{3}$$

where L_i is the number of links between all neighbors of i and k_i is the degree of node i.

Both definitions capture intuitive notions of clustering but though often in good agreement, values for C and C^{avg} can differ by an order of magnitude for some networks. Use of the global clustering coefficient has some advantages for our task. In case of Erdös–Rényi (ER) random network we can use the probability of link creation p as the mathematical expectation of the global clustering coefficient C. Moreover, this definition of the clustering coefficient is used as a primary one in [7], and if we use it, we will get more comparable results. Further, unless otherwise specified, we will use the global clustering coefficient as the measure of clustering. Nevertheless, in Table 1 the values of C^{avg} are also shown.

According to the definition of a small-world network, its clustering coefficient C_{sw} has a property: $C_{sw} \approx C_{lat}$, where C_{lat} is the clustering coefficient for the analog of the network with properties of a regular lattice. The expression for the small-world average path length follows from the second part of the definition: $\langle l_{sw} \rangle \approx \langle l_{rnd} \rangle$. Consequently, for small-world networks $\langle l_{sw} \rangle / \langle l_{rnd} \rangle \approx 1$, and from the estimation $C_{lat} \gg C_{rnd}$ we will get an estimation for clustering coefficients ratio: $C_{lat}/C_{rnd} \gg 1$. Inserting the estimations of the clustering coefficient and the average path length into the expression (1) we get the estimation for the criterion of the small-world model applicability: $\sigma_{sw} \gg 1$. By the same reasoning, we can conclude the higher the value of σ, the closer the network to the small-world state.

Values of the global clustering coefficient C and the average path length $\langle l \rangle$ for the initial UNEG network are shown in column 1 of Table 1. We can calculate the values of C_{rnd} and $\langle l_{rnd} \rangle$ for the equivalent ER random graph [5]. An ER random graph is constructed by uniquely assigning each edge to a node pair with the uniform probability. Parameters of the ER random graph generation process are the quantity of nodes N and the probability p of creating an edge. Value of p is calculated as follows:

$$p = \frac{L}{N(N-1)}, \tag{4}$$

where L is the quantity of links and N is the quantity of nodes in the network. For the UNEG $L = 642$, $N = 514$ and, consequently, $p = 0.0049$. Due to the randomness of the ER graph generation process, values of L, C, and $\langle l \rangle$ are different in each case. We repeated the generation process 100 times and calculated the mean values which are shown in column 2 of Table 1.

The mean value of links quantity in ER random graph creation process for the UNEG is quite near to L. Mean value of the global clustering coefficient for ER network corresponding to the UNEG network is 0.0047. Since probability of link creation p in ER network is the expectation of its global clustering coefficient, the value of C is quite near to $p = 0.0049$.

However, using the ER random graph creation process could not help find the correct value of the average path length $\langle l_{rnd} \rangle$ because the networks, which were generated by the process with parameters $N = 514$ and $p = 0.0049$, are disconnected. In disconnected networks, paths between some pairs of nodes are absent, and calculation of the average path length between all node pairs of a network is incorrect. Since, in our case, each ER random graph has a giant component (a connected subgraph with the largest number of nodes), as a workaround, we can calculate the average path lengths for the giant components of ER random graphs (see values with index 'G' in column 2 of Table 1). The weak point of this solution is that in our case the giant components have only near to 90% of nodes and about 99% of links of the whole network. Hence smaller quantity of nodes and higher density of links will lead to underestimation of the value of $\langle l_{rnd} \rangle$ if the average path lengths of giant components of ER random graphs are used. Since both 'triangles' and paths of length 2 from formula (2) are very rare outside giant components of ER random graphs, generated with our parameters, the global clustering coefficient of a giant component is almost the same as of the whole ER random graph (see C values in column 2 of Table 1).

To avoid this problem we can use an alternative network generation process with random relinkage procedure. This procedure is quite similar to the relinkage process in the Watts–Strogatz model for generating small-world networks from lattice networks [23]. The main difference is that in the current procedure the connectedness of the network and the degrees of all nodes are preserved.

A basic step of the random relinkage procedure consists of a random choice of two pairs of linked nodes, disruption of the old links within each pair and creation of two new links between two nodes from different pairs. Relinkage step is rolled back if it leads to violation of network connectedness. When a random relinkage step is repeated many times it saves the degree of every node and the connectedness of the network but shuffles links between the nodes. That process leads to a random network with the same degrees of nodes as in the initial network and to the preservation of connectedness of the network.

We have created the software implementation of the random relinkage procedure, and it has been applied to the UNEG network computer model. For neutralization of the stochastic effects of the random relinkage procedure, it was applied to the UNEG network 100 times. The average values of network properties for the randomized UNEG network analogs are shown in column 3 of Table 1.

The value of the average path length for the random analog of the UNEG obtained by the random relinkage procedure is 7.35. This value is more than 10% greater than the average path length for the giant component of the ER random analog of the UNEG. As mentioned above, this is the consequence of smaller quantity of nodes and higher density of links for giant components along with the effect of different nodes degree distributions in networks which are generated by different procedures (the direction of this effect requires a separate study). Moreover, the distinction between nodes degree distributions is also the cause of significant difference in magnitude of the global clustering coefficients for the two random network generation procedures.

The values of σ criterion 1 for the UNEG are shown in Table 2. For all calculation methods of C_{rnd} and $\langle l_{rnd} \rangle$ values of σ greatly exceed 1 and may be interpreted as values of a network with small-world properties. But as shown in Telesford et al. paper [21], the generally accepted approach for the detection of small-world structure in networks has significant flaws.

4 Criticism of the Generally Accepted Small-World Network Criterion

Watts–Strogatz small-world network model is defined as a hybrid between the regular lattice (in terms of its high clustering coefficient) and the random network (in terms of its short average path length). Many succeeding small-world models, e.g., Kleinberg model [9], use the same main idea. Using the generally accepted small-world criterion σ we compare a network with only one extreme case of the small-world model the random network, and ignore another extreme case the regular lattice. Comparison with the regular lattice using the traditional criterion is commonly ignored due to the absence of a simple and widely known algorithm to generate the regular analog of a network.

Depending on the method of calculating C_{rnd} the value of ratio C/C_{rnd} for the UNEG network is in the range [19.85, 25.79] (see Table 2). In any case, this ratio has a quite big value what means that global clustering coefficient of the UNEG network is very different from a random analog of the network. But from $C \gg C_{rnd}$ and $C_{lat} \gg C_{rnd}$ it does not follow that $C \approx C_{latt}$. It means that such a big value of C/C_{rnd} does not ensure that clustering of the UNEG network has a value close to regular lattices with a similar quantity of nodes and links. Consequently, when we have a high C/C_{rnd} ratio and consequently high value of σ criterion we cannot make a correct conclusion about satisfaction of one of the two requirements from the small-world definition.

Moreover, identification of the small-world structure for a network with the help of the generally accepted criterion 1 is ambiguous because value of C_{rnd} is highly dependent on the size of the network. As mentioned above, the value of the global clustering coefficient of an ER random network could be estimated by the probability of creating an edge p

$$C_{rnd}^{ER} \approx p = \frac{2L}{N(N-1)} = \frac{\langle k \rangle}{N-1} \underset{N \gg 1}{\approx} \frac{\langle k \rangle}{N} \tag{5}$$

where $\langle k \rangle$ is the average node degree. If we consider several networks which have the same structure (first of all, the same average node degree $\langle k \rangle$ and the same global clustering coefficient C) but have different quantity of nodes N, we will get the estimation for the numerator of the fraction 1:

Table 2 σ small-world criterion calculation

	1	2	3
Random analog of the network	C/C_{rnd}^{a}	$\langle l \rangle / \langle l_{rnd} \rangle^{a}$	σ
Erdös-Rényi[G]	19.85	1.83	10.83
Relinkage process	25.79	1.62	15.89

[a] The mean values for 100 results of generation process
[G] the value for the giant component of the network

$$C/C_{rnd}^{ER} \sim N \quad (6)$$

In particular, if the considered networks are small-world networks with the value of the fraction $\langle l \rangle / \langle l_{rnd} \rangle$ close to 1, then the relation between σ criterion and the size of the network N can be estimated as:

$$\sigma = \frac{C/C_{rnd}}{\langle l \rangle / \langle l_{rnd} \rangle} \sim N \quad (7)$$

This means that the value of σ criterion for small-world networks with the same structure is proportional to the quantity of nodes in those networks. Consequently, the size of a network is crucial when we use sigma criterion to check the small-world model applicability, and this fact makes using σ criterion not reliable.

For example, the two power transmission networks considered in this paper [6] are classified as a random network and a small-world network on the basis of their values for C/C_{rnd} and $\langle l \rangle / \langle l_{rnd} \rangle$. For the first network (classified as a random network) $C/C_{rnd} = 3.34$ and $\langle l \rangle / \langle l_{rnd} \rangle = 1.46$. For the second network (classified as a small-world network) $C/C_{rnd} = 22.00$ and $\langle l \rangle / \langle l_{rnd} \rangle = 1.46$. Average degrees of nodes for both networks are almost the same and have values close to 2.6 links per node. Values of σ criterion for the networks are 2.29 and 15.04, respectively. These values confirm the classification made in [6]. However, if we take into account the cause of difference in the ratios C/C_{rnd} for the networks, then our conclusions will become more ambiguous.

For the first network the ratio is calculated on the basis of the following values: $C/C_{rnd} = 0.107/0.032 = 3.344$, and for the second network the ratio is: $C/C_{rnd} = 0.088/0.004 = 22.000$. This shows that the first network has a significantly greater global clustering coefficient than the second network, and what means that the greater ratio value for the second network is explained only by different values of the global clustering coefficients of the random analog networks. Since both networks have the same average degree of nodes, the difference in the clustering coefficients for the random networks arises from the difference in the networks size: the first network has only 84 nodes, while the second one has 769 nodes. The value of C_{rnd} depends heavily on the network size, but C_{lat} should not have that dependency. Thus, while considering C/C_{rnd} in the criterion σ for testing $C \approx C_{latt}$, we implicitly add the incorrect dependency of σ on the size of the network to the criterion. Consequently, the usage of the criterion based on the ratios C/C_{rnd} and $\langle l \rangle / \langle l_{rnd} \rangle$ for classifying networks of different sizes in the case from [6] led to ambiguous conclusions.

In Telesford's paper [21] an analysis of families of small-world networks generated by the Watts–Strogatz model was performed. It has revealed that the unilateral comparison in the traditional small-world criterion has several significant flaws. In the Watts–Strogatz model from [23] the links in a regular lattice are randomly relinked with a certain relinkage probability p. Telesford's paper shows that with the growth of probability p from 0 to 1 in networks generated by the Watts–Strogatz model the small-world criterion value σ steadily increases to the maximum value and steadily decreases after reaching the maximum. It means that the dependency of p on σ is not single-valued; therefore, one criterion value has two different interpretations in terms of the Watts–Strogatz small-world model. Moreover, the traditional criterion σ does not have any certain value interval, and for different sizes of networks maximum values of σ could differ by almost two orders of magnitude. That means that the same value of the criterion σ in different cases could mean a fundamentally different small-world status of a network.

The analysis presented above demonstrated that the interpretation of the value of the σ criterion for the UNEG is very ambiguous. Despite the large value of the σ criterion for the UNEG network we cannot say that the clustering coefficient of the network is close to a regular lattice with a similar quantity of nodes. We do not know the maximum value of σ criterion for networks with the same quantity of nodes and links as in the UNEG network. Consequently, we do not know how close the UNEG network is to a perfect small-world structure. Moreover, due to the existence of the two interpretations of σ criterion, we do not know if the UNEG network differs from the perfect small-world structure in the direction of the random network structure or in the direction of the regular lattice structure.

In Kim and Obah paper [8] there is another example of difficulties in the interpretation of σ criterion. In that research the generally accepted small-world criterion σ is used to analyze the changing topology of a power transmission grid in different scenarios of failures of power transmission lines. Kim and Obah found that σ value significantly decreased in scenarios which led to major cascade failures. These facts have been interpreted in [8] as a shift from the small-world network topology to the random network structure. But the ambiguousness of the σ criterion suggests that there can be another possible interpretation of its decrease: a shift in the direction of the regular lattice structure. Moreover, there are some signs that the second interpretation is more adequate. In particular, the decrease of the σ value considered above is caused by a significant decrease in the average path length. It is a typical consequence of the long links removal from a small-world network, and it leads to a topology shift in the direction of the regular lattice structure.

5 Latticization Algorithm and New Small-World Network Criterion

To overcome the problems of the generally accepted small-world criterion in [21] a new criterion was offered by Telesford et al. for the identification of the small-world structure in networks. It is based on the latticization algorithm described in [3, 18, 20] which is used to generate the regularized analog of a network. The new criterion uses the comparison of the clustering coefficients of the current network and its regularized analog along with the comparison of the average path lengths of the current network and its random analog. The new criterion is as follows:

$$\omega = \frac{\langle l_{rnd} \rangle}{\langle l \rangle} - \frac{C}{C_{latt}} \tag{8}$$

where C_{latt} is the clustering coefficient of the latticizated analog of the current network.

For Watts–Strogatz networks the following conditions for the fractions from Eq. 2 are met: $0 \leq \langle l_{rnd} \rangle / \langle l \rangle \leq 1$ and $0 \leq C/C_{latt} \leq 1$. It follows that for Watts–Strogatz networks and for the broad class of networks in which these conditions are met the ω criterion has a certain value interval: from -1 to 1. The criterions values close to 0 conform to the structure of a small-world network, values close to -1 conform to the structure of a regular lattice, and values close to 1 conform to the random structure of the network. Dependence of p on ω in the Watts–Strogatz small-world model is single-valued; consequently, we can uniquely identify the direction of the network differences from the perfect small-world structure. Also in [21] it was shown that values of the criterion ω are almost independent from the size of a network. Thus, the new criterion does not have the fundamental shortcomings of its predecessor. It has a certain value interval for a network: from -1 to 1, and values of the criterion are almost independent from the size of a network.

The main idea of the latticization algorithm is to repeatedly execute a relinkage procedure which is similar to the relinkage procedure in the random relinkage process. The distinguishing feature of the new variant of the relinkage procedure is that the relinkage process step is executed only if the total length of the new pair of links is greater than the total length of the old ones. The testing of this condition is shown on line 9 in the pseudocode realization of the latticization algorithm (see Fig. 2). The previous steps of the relinkage algorithm are aimed to randomly choose the two pairs of linked nodes which are suitable for cross relinkage. After choosing a correct pair of nodes the relinkage procedure starts (see line 14–19 in Fig. 2). This procedure consists of the disruption of the two old links within each pair and the creation of two new links between nodes from different pairs. The relinkage procedure is rolled back if it leads to the violation of the network connectedness.

The evaluation of links lengths (or the distances between the nodes) in the latticization algorithm is described in [3, 18, 20]; it is based on the definition of a closed one-dimensional sequence of nodes ('ring of nodes') and on the metric of links length induced by that sequence. The distance between neighboring nodes in this ring (and the length of links between these nodes) is minimal and has the value of 1, while the distance between nodes in opposite parts of the ring is maximum and has the value of $[N/2]$. This rings structure corresponds to the initial one-dimensional lattice in the Watts–Strogatz small-world model. As the result of the latticization algorithm, the network is transformed into a quasiregular network wherein the majority of links connect nearest neighbors in the one-dimensional sequence of nodes.

Unlike in the case of the Watts–Strogatz model, we do not have any information about the initial one-dimensional sequence of nodes, and due to this in the latticization algorithm ring sequences are generated randomly. The latticization procedure is repeated with different initial sequences (e.g., 100 repetitions were performed for

```
function Latticization(network, relinkTryQty, findCDTryQty)
    /* network must be connected, function returns
       latticizated network                                        */
1   for i ← 1 to relinkTryQty do /* attempts to do relinkage      */
2       nodeA, nodeB ← RandomChooseLinkedNodes(network) ;
3       startRelink ← False ;
4       for j ← 1 to findCDTryQty do /* attempts to find correct nodeC
          and nodeD for cross relinkage                             */
5           nodeC, nodeD = ChooseLinkedNodes(network, nodeA, nodeB) ;
6           if nodeA = nodeC or nodeA = nodeD or nodeB = nodeC or nodeB =
              nodeD or HasLink(network, nodeA, nodeD) or HasLink(network,
              nodeB, nodeC) then
7               continue ;   /* in case of duplicates or existence of
                 planned link make new attempt to choose nodes */
8           end
9           if Distance( nodeA, nodeB) + Distance( nodeC, nodeD) >
              Distance( nodeA, nodeD) + Distance( nodeB, nodeC) then
10              startRelink ← True ;
11              break ;        /* correct nodes have been found */
12          end
13      end
14      if startRelink then
            /* nodeA-nodeB, nodeC-nodeD => nodeA-nodeD,
               nodeB-nodeC:                                        */
15          Relink(network, nodeA, nodeB, nodeC, nodeD) ;
16          if NumberOfConnectedComponents(network) > 1 then
                /* rollback relinkage because network has become
                   disconnected                                    */
17              Relink(network, nodeA, nodeD, nodeB, nodeC) ;
18          end
19      end
20  end
21  return network
end
```

Fig. 2 Realization of the latticization algorithm in pseudocode

Table 3 ω small-world criterion calculation

	1	2	3
Algorithm of network regularization	$\langle l_{rnd}\rangle/\langle l\rangle^b$	C/C_{latt}^a	ω
Latticization	0.6161	0.4261	0.1900
Geo-latticization	0.6161	0.5833	0.0328

[a]Values given for networks with minimum total length of links (among the results of 100 repetitions of each generation process)
[b]The mean values for the random relinkage process

the UNEG), and the result with the minimum total length of links is accepted as the end result of latticization.

The latticization algorithm and the small-world criterion based on it have not been used for power transmission grids analysis yet. For the current research the implementation of the latticization algorithm has been made using Python programming language and has been applied to the UNEG computer model. The basic properties of the latticizated UNEG network are shown in column 4 of Table 1. For the UNEG network the value of the new small-world criterion ω calculated taking into account the latticizated network is equal to 0.19 (see Table 3). It is near to 0 which is typical for a small-world network; this means that the UNEG network is quite close to having small-world structure. The criterion value for the UNEG network is greater than 0 which means that this network has a random rather than a regular structure.

6 Geo-latticization Algorithm

The Latticization algorithm is universal and not targeted for application to power transmission grid or any other infrastructure network. However, the nature of infrastructure networks is quite specific. First of all, their nodes have a geographic binding. Thus building quasi-regular network for random one-dimensional sequence of nodes is not adequate in this case. Infrastructure networks have well defined two-dimensional structure of nodes and metric for links length based on geographical distance.

Widespread usage of two-dimensional modifications of Watts–Strogatz small-world model [9] makes implementation of two-dimensional modification of the algorithm particularly justified. Realizing specific of infrastructure networks in the current research we have developed and programmed a new modification of the latticization algorithm and call it geo-latticization. In the geo-latticization algorithm, geographical coordinates of nodes and geodesic distance are used for creation of metric for calculating length of link in relinkage process. For implementation of the geo-latticization algorithm only new realization of function *Distance* is needed (see line 9 in Fig. 2).

For improving geo-latticization algorithm performance random selection of node pairs in relinkage procedure was changed to selection algorithm based on a function of distance. The New algorithm chooses the second pair of nodes not accidentally, but given the additional condition: one node from the second pair must have a distance to one node from the first pair not greater than a certain value. The additional condition due to the obligatory nearness of nodes from different pairs greatly increases probability of success relinkage and increases the speed of decreasing the total links length of a network. In the improved algorithm, a geohash technique [17] is used for fast search of the nearest nodes. For implementation of this feature new realization of function *ChooseLinkedNodes* (see line 5 in Fig. 2) has been done.

The result of geo-latticization of the UNEG network is shown in Fig. 1b. In the figure, we can see that some long links have been changed to shorter ones. That replacement has been done only if it was possible without changing the degree of nodes and network connectedness violation. Because of the stochastic nature of geo-latticization process this result is not determined and network configuration in Fig. 1b is only one from many possible geo-latticization results. But results in the geo-latticization algorithm are more determined than in the latticization algorithm because in geo-latticization algorithm there is explicitly defined metric for the links length. This metric is determined by geographical coordinates of nodes instead of the randomly defined ring sequence of nodes in latticization algorithm metric.

The basic properties of the geo-latticizated UNEG network are presented in column 5 of Table 1. As for the latticization algorithm 100 repetitions of the geo-latticization were performed for the UNEG and the value for network with the minimum total length of links is accepted as the end result of latticization (in column 5 of Table 1 this values have label 'M').

The average path length for the geo-latticizated UNEG network is 16.65 and this value is much less than value 67.25 for the latticizated UNEG network. Predictably, the two-dimensional quasiregular network shows a sufficiently less average path length in comparison with quasiregular network for the one-dimensional sequence of nodes. The global clustering coefficient for geo-latticizated UNEG network is 0.1614 and this value is less than value 0.2210 for the latticizated UNEG network. That difference is not accidental and has the following explanation. Nodes in two-dimensional space have more nearest neighbors than in one-dimensional space and it makes creation of clusters with same quantity of links less probable than in one-dimensional case.

The value of the small-world criterion ω based on geo-latticization for the UNEG network is 0.03. A more accurate method, which takes into account the UNEG network geographic nature shows that this network is sufficiently closer to the perfect small-world structure than that which was found by using the latticization algorithm (see Table 3). But qualitative characteristics of the UNEG network stay the same: the UNEG network has rather a random than a regular structure. Closeness of the UNEG network to a perfect small-world structure makes it relevant to analyse reliability to cascade failure effects from a network topology point of view.

7 Long Links Analysis

Analysis of real cases of cascade failures in power transmission grids in [8] and computer modeling of cascade failures in small-world networks in [15] have revealed that in power transmission grids with small-world structure a special role in cascade failure effects belongs to long links (or shortcuts). In small-world networks, shortcuts are responsible for strong reduction in the average path length. In the Watts–Strogatz small-world model a major part of new links which emerged in the relinkage process became long links. Although a large length of a power transmission line in kilometers does not necessarily mean that this line belongs to network shortcuts, there is a strong correlation of a large length of line and the possibility of it reducing the average path length in a network. The geo-latticization process virtually rules out all long links which could be eliminated from a network. Therefore, comparison of an original network with its geo-latticizated analog could help identify long links.

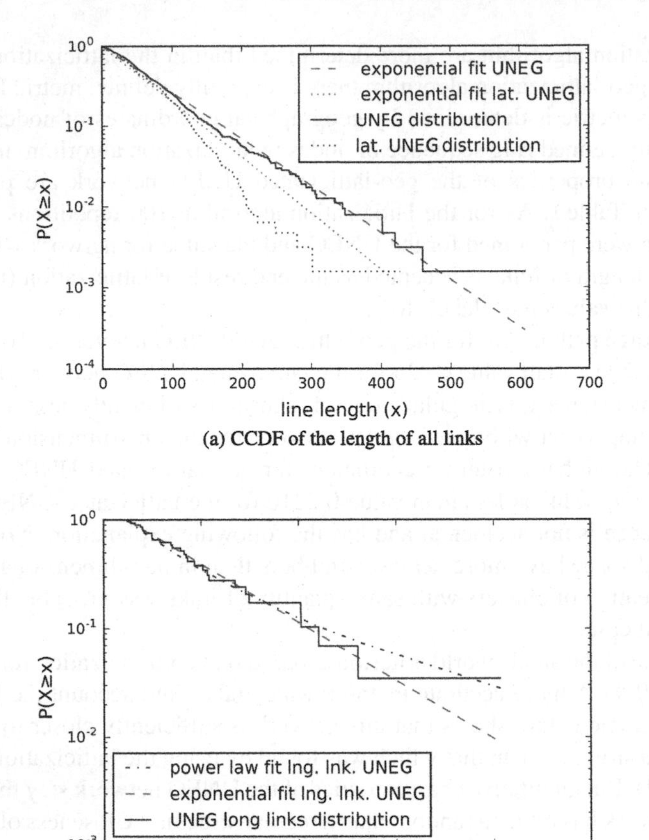

(a) CCDF of the length of all links

(b) CCDF of the length of long links

Fig. 3 The complementary cumulative distribution function (CCDF) of the length of links and fitted distribution laws: **a** for the UNEG network (distribution law: $f(x) \sim e^{-0.0133x}$) and for the geo-latticizated UNEG network (distribution law: $f(x) \sim e^{-0.0159x}$); (**b**) for long links of the UNEG network (distribution laws: $f(x) \sim e^{-0.0123x}$ and $f(x) \sim x^{-4.58}$)

In the geo-latticizated UNEG network all links except from 4 (0.4% of total quantity links) have lengths of less than 223 km but in the original UNEG network 29 have lengths greater than 223 km (see Fig. 3a). The distribution in the figure shows that in the geo-latticizated UNEG network there are many links a little shorter than 223 km. Therefore, links shorter than this limit cannot be definitely identified as shortcuts. Due to these facts, the length of 223 km has been taken as a threshold for shortcuts.

It is necessary to notice that the selected long links of the UNEG network are relatively short compared to geographical size of the UNEG network. For example, the mean of links lengths in the randomly relinked UNEG network is 914 km while the length of the longest line in the UNEG network is only 607 km. This means that the UNEG network cannot be described by the Watts–Strogatz model since it does not have long links comparable to the spatial diameter of the network.

Some other small-world models (i.e., Kleinergs model [9]) admit shorter long links but have constraints on link length distribution [4, 16]. In order to have the average path length $\langle l \rangle \sim log(N)$ (i.e., an important property of the small-world model) in a two-dimensional spatial network with randomized links (i.e., Kleinergs model), the distribution of the length of lines has to have a fat power law tail with the value of the exponent $\alpha \leq 4$ (see [4]). As seen in Fig. 3a, the empirical cumulative distribution function of the length of links for the UNEG network is approximated well by exponential distribution. The empirical cumulative distribution function of the length of shortcuts in the UNEG network can be approximated with the same quality (measured by a likelihood function) by exponential distribution or by power law distribution with the high value of the exponent $\alpha = 4.58$ (see Fig. 3b). This means that the distribution of the length of long lines does not have a sufficiently fat tail to imply slow growth in the average path length $\langle l \rangle \sim log(N)$ according to the requirements described in the Kleinergs model.

But Kleinergs model (and the majority of other spatial small-world models) is not suited for infrastructure networks such as the UNEG because many underlying assumptions of this model do not reflect the principles of their construction. Kleinergs model uses a homogeneous lattice network to construct a small-world network. But the UNEG has a spatially inhomogeneous nodes structure and a highly uneven distribution of nodes degree, and this is typical for many infrastructure networks. These features are very important for the average path length estimation because spatial areas with high density of nodes and nodes with high degree in a network make it possible to build small-world structure using relatively short shortcuts. Moreover, the estimation of dependence of the average path length from growth in number of nodes in [4, 16] assumes that linear size of a network grows as $N^{1/D}$ where D is number of dimensions in the space where network exists ($D = 2$ for the UNEG and other infrastructure networks). However, growth of real infrastructure networks is achieved not only by extending their area but also by increasing spatial density of nodes. It means that the effect from a relatively short shortcut could increase with the growth of a network.

In generally accepted spatial small-world models long links are added by randomly choosing two end points, and this is totally different from infrastructure networks. In most real cases some special (‘trunk-line’) type of links is used for shortcuts (e.g., lines of extra-high voltage (EHV) in power transmission grids). Usually, ‘trunk-line’ links form long paths (or even grids of higher order) into infrastructure networks. In such a way several relatively short shortcuts could form a path almost equivalent to a long shortcut.

In Fig. 4 diameter of a node is proportional to the total length of two longest links incident to it. This way a large diameter of a node means that it is on a path of two shortcuts. There are many large diameter nodes in the figure and almost all of them have the voltage of 500 kV or 750 kV (i.e., belong to EHV class). Moreover, from the figure we can see that in the UNEG there are several long paths formed by long

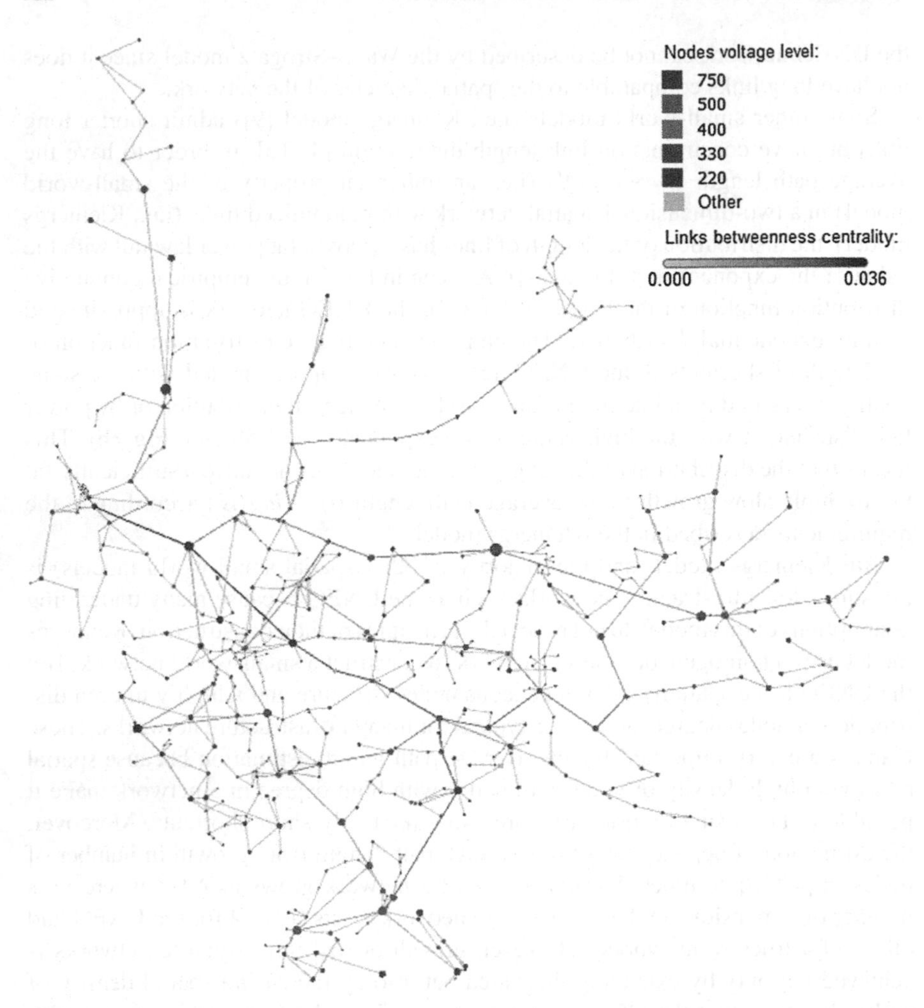

Fig. 4 Visualization of the UNEG network model. Size of a node is proportional to the total length of two longest links incident to the node (large size of a node indicates that it is on the path between two long links); color of a node is defined by voltage level of the electricity transformer in the node (see the legend, values in kV); colors of links are defined by their betweenness centrality (see color map in the legend)

links. This example confirms what was said above about 'trunk-line' links structures in infrastructure networks.

The arguments presented above show that due to the inadequacy of Kleinergs model for the UNEG case, the requirements for the distribution of links length [4, 16] must be refined using a more adequate network model. Moreover, comparison between the UNEG network and its geo-latticizated analog (see Table 1) shows that elimination of long links increases the average path length $\langle l \rangle$ from 11.93 to 16.65

hops. This confirms the high role of long links (representing only 2% of the total quantity of links) in the UNEG topology which is typical for small-world networks.

Identification of long links in the UNEG network is very important for practical UNEG operations due to the closeness of this network to a small-world structure. Identification of long links in the UNEG network will show the power transmission lines and power substations which must be points of special attention in preventing large cascade failures.

From small-world model analysis in papers [1, 2], we know that nodes incident to shortcuts have a high value of betweenness centrality. This property could be used as an alternative method for shortcuts identification. Betweenness centrality is defined as follows:

$$g(u) = \sum_{s \neq t \neq u} \frac{\sigma_{st}(u)}{\sigma_{st}} \qquad (9)$$

where $\sigma_{st}(u)$ is the number of shortest paths between nodes s and t passing through node u and σ_{st} is the number of all shortest paths between nodes s and t. A similar definition exists for betweenness centrality of links. High betweenness centrality of a node indicates its importance as a transit node in shortest paths between different pairs of nodes. Particularly, in the case of removing nodes with high betweenness centrality from a network, shortest paths between some pairs of nodes become longer and the average path length of the network increases.

Betweenness centrality was calculated for all nodes of the UNEG network computer model; the results are visualized in Fig. 5. From the figure we can see that in the UNEG network there are a few nodes of very high betweenness centrality. A matter of special interest is the fact that all those nodes are linked into one chain that runs through the center of the UNEG. The values of betweenness centrality for the nodes from the chain are in range from 0.19 to 0.35. The average value of betweenness centrality for the UNEG nodes is 0.021, the median value is 0.004, the value of 95th percentile is 0.112 and the value of 98th percentile is 0.188. Thus, the betweenness centrality values for the nodes from the chain are within 2.SS

Visual analysis of Fig. 5 can explain the special role of the nodes with a high betweenness centrality from the chain: this chain of nodes provides a topologically short path through the central part of the UNEG network. This result is achieved by large lengths of power transmission lines in the chain. It is especially important due to high density of nodes and consequently a relatively small length of power transmission lines in the central part of the UNEG network. Large lengths of power transmission lines in the chain are determined by extra-high voltage 500 and 750 kV which is economically viable for electricity transfer over such long distances.

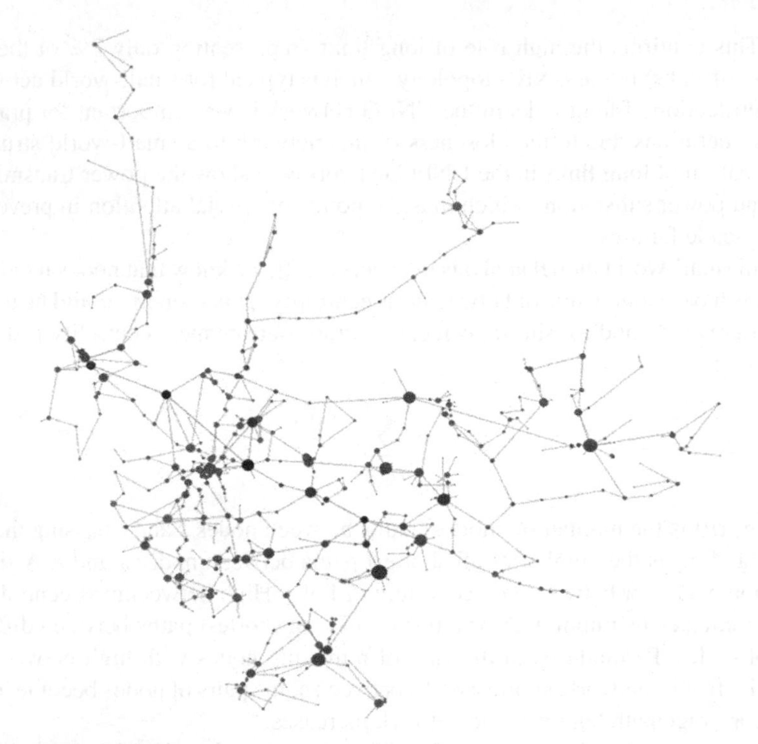

Fig. 5 Visualization of the UNEG network model. Sizes of nodes are defined by their degree; colors of nodes are defined by their betweenness centrality (*blue* nodes have the maximum centrality value)

8 Conclusion

In this paper, new methods for infrastructure networks analysis have been developed. The latticization algorithm and a new small-world criterion based on it have been used for power transmission grid analysis for the first time. The geo-latticization algorithm has been developed for a more accurate analysis of networks with geographical reference of nodes. This method helps to more accurately identify small-world properties in power transmission grids and in other infrastructure networks with geographical reference of nodes.

In this paper, complex network theory has been used to analyze the spatial and topological structure of the Unified National Electricity Grid for the first time. The new methods described above have been applied to the UNEG network. Through the use of these methods a reliable conclusion that the small-world model is applicable to the UNEG network has been made. Consequently, we have proved the necessity to conduct an analysis on the UNEG vulnerability to cascade failures with taking into account the networks topology features.

Key nodes and links which determine the small-world structure of the UNEG network have been revealed. Identification of the key power transmission lines is critical to control the reliability of the UNEG network and must be a point of special attention in preventing cascade failures. This research also discovers that the key UNEG nodes and links are combined into one chain which passes through the central part of the UNEG and transforms the whole network into a topologically compact structure.

References

1. Barrat, A., Barthelemy, M., Vespignani, A.: The effects of spatial constraints on the evolution of weighted complex networks. J. Stat. Mech. p. 503 (2005)
2. Barthelemy, M.: Spatial networks. Condens. Matter. Stat. Mech. Phys. Rep. **499**, 1–101 (2011). arXiv:1010.0302
3. Brain Connectivity Toolbox. http://www.brain-connectivity-toolbox.net (2016). Accessed 31 Aug 2016
4. Caretta Cartozo, C., De Los Rios, P.: Extended navigability of small world networks: exact results and new insights. Phys. Rev. Lett. **102**, 238703 (2009)
5. Erdös, P., Rényi, A.: On random graphs I. Publ. Math. Debrecen **6**, 290–297 (1959)
6. Han, P., Ding, M.: Analysis of cascading failures in small-world power grid. Int. J. Energy Sci. IJES **1**(2), 99104 (2011)
7. Humphries, M.D., Gurney, K.: Network small-world-ness: a quantitative method for determining canonical network equivalence. PLoS One **3**, e0002051 (2008)
8. Kim, C.J., Obah, O.B.: Vulnerability assessment of power grid using graph topological indices. Int. J. Emerg. Electr. Power Syst. **8**(6) (2007) (Article 4)
9. Kleinberg, J.M.: Navigation in a small world. Nature **406**, 845 (2000)
10. NetworkX. https://networkx.github.io (2016). Accessed 31 Aug 2016
11. OpenStreetMap. http://www.openstreetmap.org (2016). Accessed 31 Aug 2016
12. Order of the Ministry of Energy of Russia: Shema i programma razvitiya ENES na 2013–2019 godi (Scheme and development program of the UNES on 2013–2019 years). Order of the Ministry of Energy of Russia from 19.06.2013 309 (2013)
13. Pagani, G.A., Aiello, M.: The power grid as a complex network: a survey. Phys. A Stat. Mecha. Appl. **392**(11) (2011)
14. Pandit, S.A., Amritkar, R.E.: Random spread on the family of small-world networks. Phys. Rev. E **63** (2001)
15. Pepyne, J.: Topology and cascading line outages in power grids. J. Syst. Sci. Syst. Eng. **16**(2) (2007). doi:10.1007/s11518-007-5044-8 (Systems Engineering Society of China & Springer)
16. Petermann, T., De Los Rios, P.: Physical realizability of small-world networks. Phys. Rev. E **73**, 026114 (2006)
17. Python module to decode/encode Geohashes to/from latitude and longitude. Available at: https://github.com/vinsci/geohash/ (2016). Accessed 31 Aug 2016
18. Rubinov, M., Sporns, O.: Complex network measures of brain connectivity: uses and interpretations. Neuroimage **52**, 1059–1069 (2010)
19. Services for technological connection: power distribution centers. http://portaltp.fsk-ees.ru/sections/Map/map.jsp (2016). Accessed 31 Aug 2016
20. Sporns, O., Zwi, J.: The small world of the cerebral cortex. Neuroinformatics **2**, 145–162 (2004)
21. Telesford, Q.K., Joyce, K.E., Hayasaka, S., Burdette, J.H., Laurienti, P.J.: The ubiquity of small-world networks. Brain Connect **1**(5), 367–375 (2011)
22. Watts, D.J.: Small Worlds: The Dynamics of Networks between Order and Randomness. Princeton University Press, Princeton, NJ, USA (2003)
23. Watts, D.J., Strogatz, S.: Collective dynamics of small-world networks. Nature **393**, 440–442 (1998)

A New Approach to Network Decomposition Problems

Alexander Rubchinsky

Abstract A new approach to network decomposition problems (and, hence, to classification problems, presented in network form) is suggested. Opposite to the conventional approach, consisting in construction of one, "the most correct" decomposition (classification), the suggested approach is focused on the construction of a family of classifications. Based on this family, two numerical indices are introduced and calculated. The suggested indices describe the complexity of the initial classification problem as whole. The expedience and applicability of the elaborated approach are illustrated by two well-known and important cases: political voting body and stock market. In both cases, the presented results cannot be obtained by other known methods. It confirms the perspectives of the suggested approach.

1 Introduction

The most conventional statement of network decomposition problems consists in its partition into several subnetworks. Typically, it is supposed that connections within these subnetworks are significantly strongly than connections between them. The same concerns many automatic classification problems often presented as network decomposition ones. The above-mentioned partition into a small number of clearly distinct subnetworks or clusters is a result of investigation by itself. The partition allows us to formulate reasonable hypotheses about a considered system, to select a few important parameters, and so on—in brief, to understand, "what is the world in this location." Especially, it is important for many socio-economic systems, whose functioning is determined by human behavior and, therefore, cannot be adequately presented by a functional or stochastic dependence.

Numerous successful examples of decomposition/classification approach application for various system investigations are well known and comprehensively described

A. Rubchinsky (✉)
National Research University "Higher School of Economics"
National Research Technological University "MISIS", Moscow, Russia
e-mail: arubchinsky@yahoo.com

V.A. Kalyagin et al. (eds.), *Models, Algorithms, and Technologies for Network Analysis*, Springer Proceedings in Mathematics & Statistics 197,
DOI 10.1007/978-3-319-56829-4_10

in several monographs and reviews (see, for instance, [3, 4]). The result usually consists in the construction of a single, the most reasonable, in some sense, classification of the initial set. However, the experience in investigation of both model and real systems, leads to conclusions revealing other possibilities of the decomposition/classification approach. The essence of the matter consists in the construction of a ***family*** of classifications, which can include as a particular case a ***single*** classification. However, in many complicated and important situations the mentioned family includes several (sometimes many) classifications. Moreover, in such cases ***classifications themselves, forming the above-mentioned family, are of little interest***. It turned out that it is much more expedient to focus our attention on calculation of special numerical indices based on these families. The suggested indices characterize the complexity of the classification problem (CP for brevity) as whole rather than the complexity of single classifications. These indices have different meaningful interpretations in different situations, but generally they describe complexity, entanglement, perplexity, and other hardly defined, though important, properties of various systems. This approach is ***the new one***. It is necessary to underline that it does not concern the properties of single decompositions but only properties of the whole family. The exact notions and definitions require the corresponding new materials. These notions are central in the framework of the suggested approach and they are comprehensively presented in Sect. 3. They cannot be shortly formally explained in the introduction.

Construction of family of decomposition is based on the special algorithm, including some conventional stages, for instance, the original algorithm of graph dichotomy. This algorithm is presented in recent working papers [6, 7]. The first one [6] includes the comparison with the most known classification approach for the same examples (pp. 39–43). Both preprints are available in Internet at the address: www.hse.ru/en/org/hse/wp/wp7en. However, the material of the presented article mostly concerns the analysis of family of decompositions that was not considered before.

The goal of the article consists in presentation of the suggested decomposition approach and demonstration of its possibilities in analyzing of some important real systems. The material is structured as follows.

1. Introduction.
2. Classification family construction (CFC for brevity).
3. Formal definition of CP complexity.
4. Analysis of activity of the second, the third and the fourth Russian Dumas (Parliaments) in 1996–2007 years.
5. Analysis of S&P-500 stock market behavior during 2001–2010 years.
6. Conclusion.

Materials, related to Sects. 2–4, are partly presented in the recent preprint [7]. Material from Sect. 5 concerns very important, difficult and unsolved problem of short-term prediction of crises in stock markets, based on the ***share prices at some period prior to the crisis***. In spite of limited and incomplete character of obtained results, it seems that they present the first step in the right direction.

2 Classifications Family Construction

In the further described algorithm initial data about objects' proximity are presented in the well-known form of dissimilarity matrix. This means that all the objects are ordered by indices from 1 to N and for two arbitrary indices i and j numbers d_{ij}, interpreted as the degree of dissimilarity or the distance between ith and jth objects, are given. It is assumed that dissimilarity matrix $D = (d_{ij})$ $(i, j = 1, \ldots, N)$ is a symmetrical one; by definition, $d_{ii} = 0$ $(i = 1, \ldots, N)$.

Let us give the concise description of the suggested essential algorithm of CFC. At the ***preliminary stage*** the neighborhood graph G is constructed (see Sect. 2.1), basing on dissimilarity matrix D. At the ***main stage*** the formal object—neighborhood graph—is used as input.

The algorithm of the main stage is determined as a three-level procedure.

The new algorithm of graph dichotomy (see Sect. 2.2) presents the ***internal level*** of the suggested classification algorithm of the general three-level procedure of the main stage.

A special ***Divisive-Agglomerative Algorithm(DAA)***, based on the above-mentioned algorithm of graph dichotomy, is the ***intermediate level*** of the main stage. DAA produces one family of classifications (see Sect. 2.3). Pay attention that some classifications of the constructed family can coincide to one another.

At the ***external level*** (Sect. 2.4) several runs of the algorithm of the intermediate level are completed. Every run of DAA determines a family of classifications. The union (over all these runs) of all the constructed families forms the required family of classifications.

2.1 *Preliminary Stage—Neighborhood Graph Construction*

This notion is well known (see, for instance, [2]). Graph vertices are in one-to-one correspondence to given objects. For every object (say, a) all the other vertices are ordered as follows: the distance between ith object in the list and object a is a nondecreasing function of index i. All the distances are presented in dissimilarity matrix D. The first four vertices in this list and all the other vertices (if they exist), whose distance from a are equal to the distance from a to the fourth vertex in the list, are connected by edge to the vertex, corresponding to object a. It is easy to see that the constructed graph does not depend upon a specific numeration, satisfying the above conditions.

2.2 *Algorithm of Graph Dichotomy*

The input of the algorithm is an undirected graph G. There are two integer algorithm parameters:

- maximal initial value f of edge frequency;
- number of repetition T for statistics justification.

0.1. Find connectivity components of given graph G (by any standard algorithm).

0.2. If the number of components is more than 1 (i.e., graph G is disconnected), then the component with the maximal number of vertices is declared as the first part of the constructed dichotomy of the initial graph; all the other components form its second part; otherwise, go to the next step 1.

1. Preliminary stage. Frequencies in all the edges are initialized by integer numbers uniformly distributed on the segment $[0, f - 1]$.

2. Cumulative stage. The operations of steps 2.1–2.3 are repeated T times:

2.1. Random choice of a pair of vertices of graph G.

2.2. Construction of a minimal path (connecting the two chosen vertices, whose longest edge is the shortest one among all such paths) by Dijkstra algorithm. The length of an edge is its current frequency.

2.3. Frequencies modification. 1-s are added to frequencies of all edges belonging to the path found at the previous step 2.2.

3. Final stage.

3.1. The maximal (after T repetitions) value of frequency f_{max} in edges is saved.

3.2. The operations of steps 2.1–2.3 are executed once.

3.3. The new maximal value of frequency f_{mod} in edges is determined.

3.4. If $f_{mod} = f_{max}$, go to step 3.2; otherwise, go to the next step 3.5.

3.5. Deduct one from frequencies in all edges forming the last found path.

3.6. Remove all the edges, in which frequency is equal to f_{max}.

3.7. Find connectivity components of the modified graph. The component with the maximal number of vertices is declared as the first part of the constructed dichotomy of the initial graph; all the other components form second part. After that all the edges, removed at step 3.6, are returned into both graphs, except the edges, connecting vertices from different parts of the dichotomy.

2.3 *Intermediate Level—DAA*

This subsection is devoted to DAA description. Its flowchart is shown in Fig. 1. The neighborhood graph (see Sect. 2.1) is the input of DAA. Its output will be defined later. The only parameter of DAA is the maximal number k of successive dichotomies. The DAA itself consists in alternation of divisive and agglomerative stages.

At the beginning, the dichotomy algorithm (see Sect. 2.2) divides the initial (neighborhood) graph into two parts. Let us denote the found classification into two classes as D_2. Thereafter one of these two subgraphs, whose number of vertices is larger, is divided by the same algorithm into two parts that results in classification D_3 of the initial set into three classes. Classifications D_2 and D_3 are named the ***essential*** ones.

Denote them as C_2^2 and C_3^3. After entering the next essential classification D_j ($j \geq 3$) to the agglomerative stage the following operations are completed.

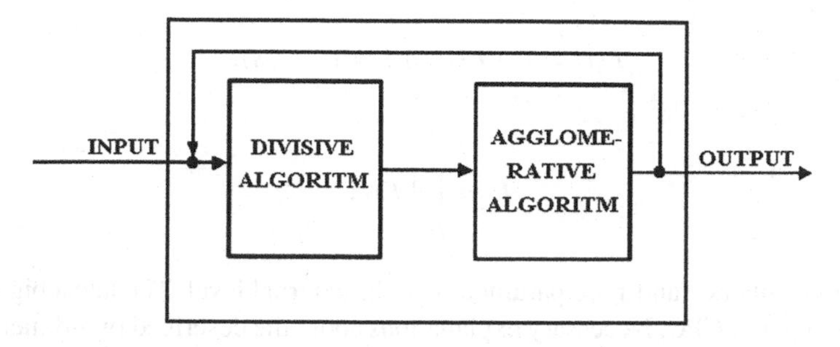

Fig. 1 DAA flowchart

Classification D_j into j classes determines the subfamily of classification into j classes (D_j itself), into $j-1$ classes (obtained by the union of subgraphs, connected by the maximal number of edges), and so on, in correspondence to the convenient agglomeration scheme (successively joining subsets, connected by the maximal number of edges), till to classification into two classes. Denote the constructed classifications as $C_j^j, C_{j-1}^j, \ldots, C_2^j$. By the construction, C_j^j coincides with D_j. Classifications $C_{j-1}^j, \ldots, C_2^j$ are named the ***adjoined*** ones.

Let us come back to the divisive stage. Among all the classes of the last constructed classification D_j select the class whose graph contains the maximal number of vertices. Delete it into two parts by the above described dichotomy algorithm. Together with other classes of D_j (except the divided one) these two classes form new essential classification D_{j+1} into $j+1$ classes. Return another time to agglomeration stage and determine adjoined classifications $C_j^{j+1}, \ldots, C_2^{j+1}$. Repeating the described steps k times produces the following family of classification:

$$C_2^2;\, C_2^3, C_3^3;\, C_2^4\, C_3^4,\, C_4^4; \ldots; C_2^{k+1},\, C_3^{k+1}, \ldots, C_{k+1}^{k+1} \tag{1}$$

This family is defined as the output of DAA. Pay attention that some classifications from list (1) can coincide to one another. The general number of classification in list (1) is equal to $\frac{(k+1)k}{2}$.

2.4 External Level—Repetitive DAA Runs

At the external level DAA is applied to the same initial graph. There are two cycles of the runs—interior and exterior. At the interior cycle DAA runs r times. It produces r families consisting of $\frac{(k+1)k}{2}$ classifications each. The exterior cycle consists of s runs of the described above interior cycle. Therefore, it produces a family, consisting of $s \times r \times \frac{(k+1)k}{2}$ classifications. Denote family of classifications, constructed on ith iteration of the exterior cycle, and on jth iteration of the interior cycle as $F(i,j)$ ($i = 1, \ldots, s; j = 1, \ldots, r$). Assume

$$F(i) = \bigcup_{j=1}^{r} F(i,j)(i = 1, \ldots, s), \tag{2}$$

$$F = \bigcup_{i=1}^{s} F(i). \tag{3}$$

Counting numbers r and s are parameters of the external level. The latest big family F is the output of CFC. Necessary explanations about the described two-dimensional structure of the constructed family are presented in the next Sect. 3.

3 Complexity Indices of AC Problems

Let us define two indices describing (in some sense) complexity of any AC problem. The definition requires two-dimensional family F, whose construction and structure are defined in the Sect. 2.4.

Index 1. Assume

$$Q(i) = d/M(i = 1, \ldots, s), \tag{4}$$

where M is equal to $r \times \frac{(k+1)k}{2}$ (general number of classifications in family $F(i)$), d is equal to the number of different classification among all the classifications in the family $F(i)$;

$$Q = \frac{1}{s} \times \sum_{i=1}^{s} Q(i). \tag{5}$$

Thus, Q is the average of complexities $Q(i)$ that are calculated separately for each family $F(i)$ $(i = 1, \ldots, s)$. By the construction, $0 < Q(i) \leq 1$ that implies the analogous inequality for value Q. This number is defined as index 1 of complexity of AC problem.

Index 2. It is defined as follows. Consider family $F(i)$ $(i = 1, \ldots, s)$. Assume that in this family classification c_p encounters m_p times $(p = 1, \ldots, t)$, where $\sum_{p=1}^{t} m_p = M$ (remember that $M = r \times \frac{(k+1)k}{2}$). Define

$$E(i) = -\sum_{p=1}^{t} \mu_p \ln(\mu_p), \text{ where } \mu_p = m_p/M \ (i = 1, \ldots, s). \tag{6}$$

$E(i)$ is the conventional entropy of division of finite family F into subsets, consisting of coinciding classifications. It is obvious that the minimal possible value of $F(i)$ is 1;

$$E = \frac{1}{s} \times \sum_{i=1}^{s} E(i). \tag{7}$$

Thus, E is the average of complexities $E(i)$ that are calculated separately for each family $F(i)$ ($i = 1, \ldots, s$). By the construction, $1 \leq E(i)$ that implies the analogous inequality for value E. This number is defined as index 2 of complexity of AC problem.

Both indices describe initial problem perplexity, entanglement, and other hardly defined but important properties of any AC problem. They are especially helpful in analysis of system dynamics.

It is intuitively clear that small (close to 0 and to 1) values Q and E correspond to relatively simple AC problems. In these problems only $k - 1$ classifications, obtained by successive divisions of the initial sets into 2, 3, …, k parts, are different. Unions in agglomerative stages and DAA runs under different initializations of a random generator do not add new classifications. Larger (close to the maximal possible) values Q and E correspond to relatively complex problems, in which found classifications essentially depend upon random generator initialization, and adjoined agglomerations differ from essential ones (see Sect. 2.3).

Let us describe the reason of using two-dimensional scheme for calculation both indices. Indices $Q(i)$ and $E(i)$ depend on parameter r (number of DAA repetitions). In the most important cases they do not have a limit as functions of r when r tends to infinity or they have different limits depending on i ($i = 1, \ldots, s$). Therefore, in order to find some stable answer it is possible to use averaging over i as it is done in formulae (5) and (7). Computational experiments have demonstrated that values Q and E (of course, also depending on r) show stability (at least, technical one) under increase of r for relatively small s. These circumstances justify the suggested approach to complexity definition.

In this connection it is possible to remember (as an analog) thoroughly considered in chapter III of the famous book [1] example of random walk. The following game is studied. If an unbiased coin falls (after tossing) heads up, player B pays to player A \$1; otherwise, player A pays to player B \$1.

Denote the gain of player A for first k such games as Z_k. Of course, Z_k can be positive, negative or 0. Denote by g_n^+ the number of values k between 0 and n, such that $Z_k \geq 0$, by g_n^-—the number of values k between 0 and n, such that $Z_k \leq 0$. Intuitively it seems that $\lim_{n\to\infty} g_n^+/g_n^- = 1$. Yet in [1] it is proved that in this case intuition is wrong, and the above-mentioned limit simply does not exist. It seems that it contradicts to the symmetry of the game. But the symmetry (existence of the limit equal to 1) is rebuilt if one considers simultaneous accomplishing of large number of such games. It is possible to say that arbitrary long sequence of single games does not converge, while shorter sequence of long sequences converges.

Of course, in the considered in the present work case mathematical essence of the absence of convergence in separate sequence is more complicated. The original cause consists in the suggested algorithm of dichotomy. It can produce significantly different divisions into two classes under arbitrary number of random paths (see, for instance, Fig. 3 in preprint [7]). It is not a mistake but the kernel of the suggested approach that constructs a family of classifications, generally differing one to another.

It is easy to see directly from the definitions that both complexity indices depend upon all the three parameters k, r and s. Theoretically, important question about exact definitions of indices that do not depend upon these parameters remains open. Table 6–8 from preprint [7] illustrate change of index Q for small values r and k under $s = 1$. Some experimental facts, concerning index E, are considered further in Sect. 5.

In the following material, the case in point does not relate to construction of the adequate model of socio-economic systems. I believe that in any complicated situations, connected to human behavior, it is almost impossible. The matter concerns the suggested ***methods of analysis of exact data*** describing the activity of such systems: votes, shares cost, and so on, without any assumptions and hypotheses about human behavior producing just the given results.

Thus, the case in point is not about models but about data analysis algorithms. Therefore, the justification of choice of one or other algorithm parameters is not presented as well as justification of choice of algorithms themselves. By contrast with natural science verification in social science is crucially impossible—we cannot abolish results of voting in parliaments or trading in stock markets and ask the participants to do the same another time. The only reliable thing consists in common sense and experience of specialists those are seriously engaged in interpretation of presented numerical results.

4 Analysis of Voting in RF Duma (Parliament)

The suggested approach to calculation complexity of AC problems was applied to analysis of voting in second, third, and fourth RF Duma (1996–2007). For every separate month of the considered period all the votes are considered. To every ith deputy ($i = 1, 2, \ldots, m$) a vector $v_i = (v_1^i, v_2^i, \ldots, v_n^i)$ is related, where n is the number of votes in a given month. Note, that the number m of deputies, though slightly, changed from period to period. Of course, at every moment the number of deputies is always equal to 450. Yet, during 4 years some deputies dropped out while the other ones came instead. The number of deputies participated in Duma voting activity in 1996–1997 was equal to 465, in 1998–1999—to 485, in 2000–2003—to 479 and in 2004–2007—to 477.

Assume

$$v_j^i = \begin{cases} 1, & \text{if ith deputy voted for jth proposition;} \\ -1, & \text{if ith deputy voted against jth proposition;} \\ 0, & \text{otherwise (abstained or not participated).} \end{cases}$$

Dissimilarity d_{st} between sth and tth deputies is defined as usual Euclidian distance between vectors v_s and v_t. The dissimilarity matrix $D = (d_{st})$ is the initial one for finding deputies classifications by the method, described in Sect. 2. Index Q, described above, is considered as the measure of complexity.

The following Table 1 presents the complexity of corresponding classifications for every month of the voting activity of second, third, and fourth RF Duma. The numbers in the first column are the dates (year and month). The numbers in the second column are equal to the number of votes in the corresponding months. Numbers in the third columns are equal to complexity of the corresponding AC problem, calculated following the definition of this notion in Sect. 3. Here the number k of consecutive dichotomies is equal to 10, the number r of DAA runs also is equal to 10, so that the maximal number $\frac{(k+1)*k}{2} * r$ of classifications is equal to 550. Empty rows correspond to months without any voting activity.

The numbers in the third column in Table 1, i.e., complexity of classifications based on the voting results, demonstrate noticeable variability, though some trend are seen at once, by "unaided eye." Smoothed data, i.e., average value for half years, thereafter for years, and, finally, for whole period of every Duma activity, are presented in Table 2.

It seems that low value of complexity in 2002 was due to creation of party "United Russia" and connected with attempts of straightening out the activity of Duma. It is surprising—at first sight—that in the forth Duma in the condition of constitutional majority of this party the level of complexity is noticeably higher than in the third Duma (0,235 opposite to 0,147), in which no party had majority.

Conclusions of such a type were not made analyzing Duma activity for the same period by other methods. It is possible to say that for voting political bodies high complexity of corresponding AC problems means inconsistence, maladjustment, irrationality of the whole body rather than of single fractions and deputies.

5 Stock Market Analysis

The stock market S&P-500 (500 greatest companies in USA) is considered. The distance between two shares is determined as follows.

1. Let us define the basic minimal period, consisting of l consecutive days. All the data found for the period $x, x-1, \ldots, x-l+1$ are related to day x. Assume the length l of considered period equal to 16. This choice is determined by the following meaningful reasons: for short period, data are too variable, for long period—too

Table 1 Complexity of voting

Complexity of voting generated classifications in second Duma (1996–1999)											
1	2	3	1	2	3	1	2	3	1	2	3
9601	174	0.610909	9701	234	0.456364	9801	248	0.421818	9901	416	0.207273
9602	321	0.625455	9702	427	0.445455	9802	366	0.330909	9902	354	0.250909
9603	295	0.581818	9703	334	0.381818	9803	347	0.469091	9903	482	0.369091
9604	470	0.683636	9704	437	0.316364	9804	334	0.436364	9904	384	0.372727
9605	263	0.938182	9705	169	0.485455	9805	292	0.398182	9905	228	0.449091
9606	269	0.827273	9706	762	0.238182	9806	489	0.534545	9906	768	0.392727
9607	450	0.263636	9707			9807	493	0.352727	9907		
9608			9708			9808			9908		
9609			9709	337	0.201818	9809	405	0.390909	9909	292	0.241818
9610	432	0.494545	9710	354	0.247273	9810	326	0.507273	9910	338	0.270909
9611	226	0.567273	9711	253	0.289091	9811	338	0.327273	9911	696	0.218182
9612	566	0.465455	9712	530	0.265455	9812	534	0.392727	9912	243	0.430909
Complexity of voting generated classifications in third Duma (2000–2003)											
1	2	3	1	2	3	1	2	3	1	2	3
0001	71	0.547273	0101	141	0.109091	0201	279	0.183636	0301	144	0.203636
0002	228	0.112727	0102	254	0.245455	0202	380	0.063636	0302	350	0.136364
0003	177	0.387273	0103	268	0.085454	0203	311	0.081818	0303	382	0.160000
0004	368	0.112727	0104	409	0.187273	0204	640	0.114545	0304	519	0.136364
0005	279	0.141818	0105	248	0.296364	0205	353	0.138182	0305	248	0.141818
0006	454	0.149091	0106	683	0.069091	0206	956	0.072727	0306	677	0.083636
0007	301	0.078182	0107	825	0.132727	0207			0307		

(continued)

Table 1 (continued)

Complexity of voting generated classifications in third Duma (2000–2003)											
0008			0108			0208			0308		
0009	144	0.154545	0109	200	0.140000	0209	329	0.120000	0309	208	0.221818
0010	371	0.169091	0110	360	0.069091	0210	541	0.067273	0310	428	0.072727
0011	240	0.103636	0111	668	0.160000	0211	448	0.065454	0311	400	0.203636
0012	483	0.138182	0112	600	0.101818	0212	531	0.058182	0312		
Complexity of voting generated classifications in fourth Duma (2004–2007)											
1	2	3	1	2	3	1	2	3	1	2	3
0401	101	0.360000	0501	130	0.283636	0601	168	0.216364	0701	243	0.214545
0402	220	0.101818	0502	209	0.421818	0602	204	0.289091	0702	189	0.356364
0403	270	0.141818	0503	237	0.225455	0603	256	0.265455	0703	262	0.123636
0404	295	0.101818	0504	355	0.090909	0604	255	0.147273	0704	368	0.187273
0405	249	0.325455	0505	255	0.123636	0605	179	0.194545	0705	190	0.118182
0406	385	0.143636	0506	300	0.338182	0606	365	0.085454	0706	448	0.169091
0407	378	0.372727	0507	240	0.141818	0607	260	0.221818	0707	320	0.310909
0408	268	0.303636	0508			0608			0708		
0409	101	0.274545	0509	174	0.325455	0609	230	0.114545	0709	141	0.167273
0410	252	0.261818	0510	266	0.360000	0610	305	0.278182	0710	350	0.298182
0411	355	0.349091	0511	359	0.232727	0611	528	0.320000	0711	337	0.227273
0412	535	0.250909	0512	426	0.225455	0612	463	0.260000	0712		

Table 2 Smoothed complexity data

	Half 1	Half 2	Half 3	Half 4	Half 5	Half 6	Half 7	Half 8
Duma 2	0.711	0.448	0.387	0.251	0.432	0.394	0.340	0.290
Duma 3	0.242	0.129	0.165	0.121	0.109	0.078	0.144	0.166
Duma 4	0.196	0.302	0.247	0.257	0.199	0.239	0.195	0.251
			1st year	2nd year	3rd year	4th year		
		Duma 2	0.606	0.332	0.415	0.320		
		Duma 3	0.190	0.145	0.096	0.151		
		Duma 4	0.249	0.252	0.217	0.217		
			Duma 2	Duma 3	Duma 4			
			0.418	0.147	0.235			

smooth. The choice of parameters in general is discussed in the last paragraph of Sect. 3.

2. Prices of all the shares at closure time are considered for days $x, x-1, \ldots, x-l+1$. The matrix R of pairwise correlation coefficients is calculated basing on these prices.

3. Distance d_{ij} between two shares (say, i and j) is defined by the formula $d_{ij} = 1-r_{ij}$, where r_{ij} is the correspondent element of matrix R.

The determined distance d is close to 0 for "very similar" shares and is close to 2 for "very dissimilar" shares. Therefore matrix $D = (d_{ij})$ is considered as the dissimilarity matrix in the AC problem, whose objects are shares related at the considered period x, $x-1$, ..., $x-l+1$ to the stock market S&P-500. Pay attention that these sets of shares can be different for different last days x of a current period.

The dissimilarity matrix defines an automatic classification problem. Let the number k of successive division is equal to 2 (see Sect. 2.3). It means that the initial set is divided into two parts, thereafter the larger of these parts also is divided into two parts and finally two parts of the three ones (connected by the maximal number of edges) are pooled to one part. Thus, three classifications—one into three classes and two into two classes—are constructed. The two latest can coincide or can be different.

Number r of runs is equal to 150. It means that family F consists of $450 = 3 \times 150$ classifications, some of which can coincide. Therefore, it is possible to calculate the complexity index 2 of family F (see formulae (6) and (7). This number (entropy of family F) is related to day x (the last day of a 16-days period. By the construction, calculation $E(x)$ requires data only for day x itself plus 15 previous days. Therefore, $E(x)$ can be calculated in the evening of day x—practically in several minutes after closure time.

However, use of random generator in the algorithm and (even in a greater extent) the complexity of the considered situation lead to the following parameters: tenfold repetitions of 150-fold calculations. The results of such tenfold repetitions of 150-fold calculations for a day x (for instance, 01.01.2001) are presented in the first row of Table 3. The other results of tenfold repetitions of 150-fold calculations are presented in the second row of Table 3. Yet average values of entropy taking over two rows of Table 3 are very close (they differ approximately in 0.002).

Let us consider the period since 01.01.2001 until 01.12.2010. This period includes two big crises: dotcom crisis in 2001 and hypothec crisis (becoming world crisis) in 2008. The entropy $E(x)$ is calculated for every day x from the considered 3652-days period. The average values of entropy at every day are stable enough, as well as in the first day of the period (see Table 3).

Entropy $E(x)$ at every day x are presented in Table 4. For commodity, results for every year are given separately. Some groups of seven consecutive days are marked by gray background. It will be explained later.

5.1 Crises Patterns

Let us begin with the following example. The values of entropy for 7 days, prior to 04.03.2001 and 22.09.2008, i.e., 5 and 7 days before big crises, are presented in Table 5. Values in the same columns of Table 5 are significantly different. Both sequences are marked by gray background in Table 4.

Denote values in the first column as x_1 and y_1, in the second column as x_2 y_2, and so on, till to values in the seventh column, denoted as x_7 and y_7. The values x_1, x_2, …, x_7 and values y_1, y_2, …, y_7 satisfy the following system of inequalities:

$$\begin{cases} z_5 > z_1, z_5 > z_2, z_5 > z_3, z_5 > z_4, z_5 > z_6, z_5 > z_7, \\ z_3 > z_1, z_3 > z_2, z_3 > z_4, z_3 > z_6, z_2 > z_7, \\ z_6 > z_7, z_5 > 6, z_3 > 5.95, z_4 < 6. \end{cases} \tag{8}$$

Inequalities from first row mean that value z_5 is greater than all the other values; inequalities from the second row mean that value z_3 is greater than all the other values, except z_5; next inequality $z_6 > z_7$ (together with inequality $z_5 > z_6$) means that three last values monotonously decrease. The three last inequalities express one-sided constraints of values z_5, $z_3$and z_4.

Consecutive seven values of entropy whose seventh value correspond to arbitrary day x can satisfy or not satisfy to system of linear inequalities (8). We see that 7-tuples that correspond to 04.03.2001 and to 22.09.2008 satisfy to system (8). These 7-tuples are marked be gray background in Table 4. Moreover, there are only three days during all the 10-year period, whose 7-tuple satisfy system (8)—except the two above mentioned cases. These cases are also marked in Table 4. Therefore, it is possible to consider system (8) as a ***pattern***, which corresponds to beginning of big crises. Anyway, absence of such a pattern means (almost exactly) that no crisis is expected in several next days. At the same time, presence of such a pattern points out to the big chance of crisis in near future. At least, it is very desirable to be ready to any surprises at stock market.

The accomplished computational experiments point out to noticeable connection between values of entropy of constructed families of classifications and behavior of

Table 3 Short sequences and their average values

Short sequence	1	2	3	4	5	6	7	8	9	10	Average
Entropy 1	5.335	5.396	5.328	5.323	5.367	5.345	5.358	5.323	5.353	5.392	5.352
Entropy 2	5.300	5.385	5.357	5.321	5.383	5.356	5.375	5.354	5.360	5.306	5.350

Table 4 Entropy at every day in 2001–2010

2001													
5.353	5.229	6.023	5.020	4.904	5.115	4.991	5.348	5.682	4.638	4.622	4.968	4.879	4.680
4.623	5.977	5.446	6.077	5.227	6.061	6.041	5.949	5.371	5.386	5.929	5.894	5.317	5.611
5.656	5.047	6.091	5.531	4.711	4.613	5.223	5.438	5.330	5.483	5.216	5.366	5.382	5.658
4.922	5.203	5.524	5.556	5.725	5.428	5.477	5.253	5.786	5.045	5.580	5.566	5.694	4.979
4.887	5.854	5.981	5.757	6.028	5.351	5.175	5.362	5.877	4.662	4.835	5.750	5.387	5.687
5.442	5.500	5.780	5.481	5.356	5.431	4.633	4.457	4.506	4.742	4.935	4.726	5.119	4.543
4.448	5.537	5.640	5.265	6.091	6.011	4.552	4.765	4.626	5.439	5.064	5.102	5.527	5.892
5.999	5.863	6.091	5.935	5.809	5.875	4.745	5.211	5.655	5.186	4.702	5.949	5.772	5.668
5.289	4.950	5.421	5.290	4.602	4.757	5.000	5.053	6.074	5.828	5.452	5.984	5.424	5.219
5.196	4.992	5.218	5.495	5.865	5.933	5.795	5.600	5.443	5.593	5.524	5.780	5.528	5.841
5.014	4.978	4.736	4.813	4.807	4.966	5.395	5.289	4.975	5.755	5.903	5.964	5.146	5.137
5.593	5.612	5.408	5.594	5.122	5.383	5.559	5.098	5.829	5.938	5.713	5.569	5.786	5.369
5.214	4.650	4.700	6.028	4.868	4.943	5.520	5.391	4.805	5.846	5.705	6.057	5.733	5.424
4.800	5.189	5.974	5.798	5.153	5.505	4.917	5.492	5.481	4.830	5.604	5.855	5.924	5.216
5.702	5.613	5.597	5.278	5.838	5.222	4.593	5.062	5.551	5.642	5.834	4.735	5.546	5.777
6.060	5.870	5.667	5.096	5.002	5.137	5.956	5.876	5.926	4.885	6.075	5.724	6.069	5.872
5.133	5.818	5.486	4.663	5.204	5.134	5.106	4.845	5.512	5.565	5.079	5.479	4.861	5.775
6.003	5.901	5.523	5.942	5.393	5.226	4.667	4.863	4.688	5.216	4.667	5.601	4.994	4.623
4.439	4.530	4.960	4.541	5.180	5.172	4.710	5.312	5.860	4.492	4.500	4.497	4.468	4.460
4.457	4.528	4.472	5.677	4.920	5.502	5.972	5.482	5.377	5.737	5.729	5.674	5.973	6.077
4.677	5.263	5.594	5.130	4.936	4.823	4.978	4.513	5.485	5.619	5.172	5.422	5.839	6.000
5.838	5.971	5.230	5.204	6.087	5.956	5.641	5.132	5.323	5.213	5.339	4.894	5.525	5.408
4.539	5.551	5.003	5.288	4.891	5.046	4.710	5.126	5.455	5.542	4.496	5.172	5.672	5.141
4.650	4.610	5.889	4.831	5.228	4.822	5.162	4.630	5.326	5.619	6.090	5.926	5.985	6.001
5.640	6.080	5.452	5.457	5.643	5.629	4.918	4.874	4.836	5.407	4.726	5.674	5.649	6.046
4.883	5.745	5.846	6.038	5.915	5.066	5.625	5.659	4.936	5.058	5.013	5.213	5.069	4.986
4.760													

(continued)

Table 4 (continued)

2002													
5.279	6.029	5.993	5.652	5.739	5.745	5.696	5.604	5.208	5.217	5.842	4.981	5.676	4.975
5.437	5.218	4.882	5.126	5.694	4.942	5.030	4.902	5.167	4.732	5.120	4.869	5.944	6.067
5.064	4.705	5.469	5.104	5.569	5.357	5.450	5.161	5.834	5.760	5.128	5.037	5.127	5.984
5.492	6.025	5.709	6.052	5.915	5.702	5.793	5.192	5.696	5.253	5.831	5.138	5.705	5.606
5.589	5.605	5.110	5.876	5.310	5.584	5.446	4.916	4.549	5.049	4.567	4.577	4.518	4.581
5.603	4.984	5.939	5.232	4.951	4.825	6.043	5.113	5.175	5.432	5.532	5.018	5.996	5.476
5.517	5.841	5.711	5.765	5.838	4.736	5.906	5.728	5.209	5.923	6.073	5.923	5.641	5.025
4.832	5.564	5.518	5.082	4.594	4.616	4.967	5.681	5.057	5.890	5.417	5.623	5.592	5.728
5.993	5.864	5.418	5.135	5.165	4.758	4.614	5.050	4.763	5.296	5.929	5.273	5.109	5.100
4.905	6.017	5.756	5.977	6.029	5.913	5.237	5.636	5.599	6.078	5.333	5.718	5.795	5.600
5.672	5.594	4.698	5.411	5.909	5.973	5.844	5.773	4.712	5.126	5.232	5.290	6.093	5.641
4.599	5.178	4.650	4.587	4.554	4.559	4.734	4.917	4.712	5.481	5.143	5.307	5.330	5.388
6.057	5.815	4.857	5.207	5.021	5.099	5.768	5.689	4.930	5.504	5.101	4.979	5.359	5.238
5.101	4.562	4.973	4.707	5.080	4.835	4.744	5.337	4.596	5.950	4.829	6.078	5.910	5.017
5.452	4.984	4.439	4.499	4.694	4.736	4.465	4.450	4.524	4.486	5.063	5.390	5.513	4.735
5.218	4.644	4.877	4.608	4.684	4.791	5.388	5.113	5.638	5.270	5.108	4.959	5.057	5.474
5.452	4.443	5.494	4.439	5.036	4.499	4.443	4.762	5.257	4.855	5.598	5.210	4.821	4.676
5.000	6.053	5.024	5.918	4.920	5.189	4.827	5.261	5.525	4.837	4.746	5.833	6.079	4.846
5.697	5.525	5.912	5.464	5.632	5.787	5.758	5.952	5.964	5.807	5.964	5.984	5.364	4.507
5.025	4.480	4.541	5.136	5.624	5.707	4.654	5.112	5.334	6.008	5.941	4.569	4.776	4.866
5.306	5.845	5.160	5.030	5.155	5.373	6.047	5.814	5.572	5.423	5.055	4.575	4.605	4.517
4.899	5.060	4.902	4.985	5.835	5.327	4.790	5.658	6.081	5.962	6.013	6.019	6.052	5.860
5.484	5.279	5.808	5.913	5.057	4.982	5.476	5.315	5.310	6.067	4.883	6.070	6.092	6.084
5.992	6.063	4.925	5.410	4.679	5.037	5.595	4.786	4.745	4.511	4.852	5.259	5.724	5.104
5.012	5.653	6.073	6.093	5.932	5.950	4.787	4.838	4.996	5.322	4.744	5.090	4.981	5.683
5.152	5.261	5.753	5.734	5.628	5.571	5.874	4.439	5.475	5.675	6.085	5.500	5.845	5.382
5.472													

(continued)

Table 4 (continued)

2003													
5.771	5.762	5.489	5.763	4.733	5.613	4.633	5.169	4.744	4.801	4.758	4.842	4.816	5.629
5.186	6.031	5.945	5.524	5.706	5.863	5.413	4.604	5.118	5.297	4.904	4.566	5.295	4.520
4.564	4.628	4.891	5.564	4.946	5.137	5.834	5.564	4.810	5.165	5.748	5.445	5.231	5.858
4.802	4.542	5.126	5.262	4.728	4.744	5.005	5.661	5.978	5.205	6.019	4.668	6.067	5.578
5.725	4.985	5.461	5.682	5.576	5.467	5.829	4.911	5.971	5.835	5.841	5.808	5.349	4.859
4.981	5.345	4.550	4.532	4.639	6.077	5.752	5.728	4.866	4.772	5.579	4.667	4.439	4.520
4.503	5.907	5.453	5.324	5.416	5.511	5.711	4.596	5.169	5.348	4.698	5.712	5.898	5.616
5.258	5.567	5.764	4.911	5.348	5.668	5.316	4.452	4.817	5.035	5.387	5.760	5.360	4.903
5.626	4.882	5.270	4.996	4.895	5.788	5.989	5.111	5.945	5.400	5.377	5.393	5.446	4.994
5.174	4.871	5.749	5.686	5.177	4.439	5.191	5.302	5.494	4.826	5.511	5.416	5.136	5.029
4.830	5.256	6.053	5.885	5.924	5.903	5.725	5.554	5.811	5.065	5.538	4.597	4.642	4.798
4.452	4.949	5.984	5.243	5.238	5.622	4.777	4.493	4.699	4.722	5.079	5.770	5.716	5.840
5.683	5.853	5.824	5.765	5.353	5.637	5.598	5.976	5.485	5.237	5.395	5.902	5.711	5.058
5.169	5.206	5.150	5.508	5.285	5.319	4.952	5.587	5.324	4.993	4.532	4.593	4.635	4.739
4.935	5.699	4.498	5.406	5.789	4.842	5.817	5.658	5.239	5.496	5.236	6.034	6.053	5.927
6.039	6.027	5.199	5.286	5.983	5.973	5.487	4.994	5.743	6.106	6.046	5.360	5.773	6.039
5.692	5.478	5.651	5.519	5.689	5.846	5.938	5.993	5.267	4.762	4.854	4.617	5.204	5.150
5.412	5.970	5.790	5.423	5.661	4.746	6.087	5.346	4.643	4.678	5.365	5.173	5.434	5.611
5.014	5.267	5.372	5.791	5.380	5.613	4.977	5.032	5.515	5.858	6.058	5.511	6.029	5.514
5.170	5.502	4.589	5.567	5.340	5.288	5.957	5.508	5.761	5.449	4.705	4.661	5.782	5.578
4.693	5.232	4.856	4.612	4.593	5.121	4.492	4.508	4.706	4.821	5.227	5.732	5.136	5.440
5.595	5.276	5.896	6.053	6.016	5.252	5.263	5.323	6.066	5.351	5.592	6.013	4.990	4.722
5.671	4.716	5.402	4.930	5.978	4.910	5.470	5.540	5.193	5.756	5.289	5.003	5.089	5.632
5.496	5.181	4.439	5.202	5.681	5.010	4.819	5.549	5.077	5.763	5.180	4.463	4.511	5.508
5.560	4.528	5.465	5.111	4.963	5.054	5.443	5.318	4.680	5.492	5.337	4.820	4.862	5.030
5.731	5.814	5.861	6.063	4.954	5.664	5.915	5.619	5.347	4.687	5.326	5.602	5.829	4.612
5.156													

(continued)

Table 4 (continued)

2004													
4.799	5.942	5.072	4.886	5.578	6.076	5.349	5.209	5.015	5.145	5.433	5.210	5.985	5.768
5.244	5.960	5.457	5.308	5.580	5.304	5.297	5.314	5.788	4.694	4.831	4.685	5.240	5.348
5.501	5.954	5.282	5.071	5.213	5.801	5.160	5.237	5.439	5.085	4.919	5.864	5.477	5.844
5.397	6.036	4.556	5.592	5.635	5.280	5.397	6.080	5.779	5.352	5.583	5.370	5.799	5.396
5.143	5.893	5.694	5.120	5.799	5.882	5.330	4.733	4.946	5.519	4.922	5.192	4.863	5.882
5.499	5.992	5.611	5.577	4.753	5.313	5.659	5.538	5.520	5.408	4.840	5.021	4.825	4.635
5.704	5.411	5.181	5.395	5.733	5.988	5.440	5.537	5.432	5.291	5.084	4.730	4.751	4.871
4.554	4.463	4.659	4.572	4.439	5.169	5.507	5.811	4.736	5.180	4.809	5.329	5.634	5.406
5.820	5.407	5.205	4.853	5.780	6.038	5.969	5.336	5.571	5.981	5.935	5.568	5.955	5.400
5.555	5.016	5.562	4.955	5.068	5.188	4.439	4.789	4.650	4.573	5.114	4.863	4.959	5.275
6.052	4.899	5.594	5.821	5.141	5.925	5.745	5.683	5.605	5.420	4.986	5.052	5.130	5.047
4.965	6.001	6.043	5.236	5.803	5.575	5.418	6.031	5.795	5.114	5.666	6.036	6.091	4.821
5.905	5.541	5.108	5.070	5.368	4.971	4.839	5.332	5.220	4.606	4.970	5.284	4.955	5.842
5.841	5.584	5.814	5.640	5.492	5.978	5.894	4.728	5.241	5.391	4.658	5.192	5.474	4.605
5.893	5.583	5.172	4.578	4.602	5.461	5.376	5.993	5.181	5.802	5.791	5.485	5.730	4.671
5.167	4.818	4.855	5.390	5.912	5.662	6.024	5.787	5.653	4.905	4.665	4.789	4.927	4.535
4.618	4.452	5.048	4.600	5.076	4.956	5.083	5.126	5.521	5.261	5.636	4.981	4.643	4.439
4.637	4.784	4.876	5.033	4.881	5.229	5.923	4.791	4.600	5.318	4.823	4.798	4.930	4.744
5.583	4.887	4.850	4.515	5.647	4.988	5.188	5.603	5.761	5.670	5.900	5.266	5.130	5.907
5.452	5.501	5.841	5.308	4.818	5.350	5.602	5.417	5.184	5.286	4.898	5.431	5.133	5.125
5.532	4.588	4.712	4.463	4.650	5.448	5.045	5.158	5.783	5.193	5.016	4.896	4.861	5.397
4.717	5.821	4.809	5.546	5.440	5.613	5.425	5.050	5.693	5.458	5.017	4.629	5.202	4.498
4.640	4.472	4.439	4.502	4.439	4.650	4.527	4.967	5.211	4.830	5.699	5.139	5.358	5.527
5.717	4.941	5.512	5.171	4.439	4.770	5.954	5.087	4.899	5.507	5.658	5.500	5.630	5.399
5.376	5.283	4.884	4.602	4.763	5.543	5.641	4.999	5.020	5.497	5.586	4.967	6.006	5.104
5.484	5.807	6.042	6.036	6.059	5.233	5.092	5.995	5.890	5.630	5.644	5.359	5.519	4.863
5.573	5.029												

(continued)

Table 4 (continued)

2005													
5.275	4.650	5.819	5.255	5.675	5.857	4.739	4.862	4.463	4.563	4.576	4.763	5.303	5.023
5.279	5.823	5.995	4.777	5.754	4.775	5.139	5.495	5.568	5.856	5.877	5.690	5.763	5.473
5.287	4.777	4.439	5.469	5.645	5.279	5.052	4.675	5.352	4.901	4.930	5.388	4.815	4.576
5.019	4.587	5.028	4.545	4.685	5.124	5.584	5.111	5.118	5.485	5.715	5.408	5.166	5.986
5.171	5.680	4.815	5.828	5.546	5.288	5.719	4.601	4.636	4.519	4.748	4.919	5.391	5.163
6.031	5.913	5.534	4.615	5.523	5.678	5.349	5.618	5.505	5.594	5.774	5.015	5.056	5.024
5.115	5.018	5.121	5.125	5.778	4.845	6.007	5.131	4.904	4.439	6.038	5.573	5.552	4.672
4.870	5.685	5.735	5.616	5.886	5.386	5.273	4.610	4.665	5.078	4.439	4.439	4.617	4.439
4.711	5.618	5.446	5.976	5.387	5.952	6.093	5.335	5.102	5.913	5.824	5.064	5.962	5.257
5.079	5.119	5.233	5.474	5.828	4.917	5.881	5.759	4.680	5.168	6.066	5.116	5.674	5.747
5.511	5.364	5.227	5.059	5.273	4.864	4.472	4.929	5.201	5.063	5.423	5.934	5.204	5.537
5.748	5.825	5.249	4.880	5.459	5.655	5.235	5.568	5.292	5.198	5.894	5.807	5.981	5.291
5.003	5.427	5.614	5.606	4.643	5.143	5.815	5.616	5.578	4.629	4.695	5.275	5.046	4.702
5.419	5.113	4.863	5.756	5.101	5.908	5.648	4.876	4.439	4.610	4.914	4.983	5.639	5.105
4.452	4.765	5.186	5.081	5.161	5.148	5.422	5.429	5.512	5.298	5.711	5.809	5.812	4.609
5.297	5.380	5.547	5.049	5.335	5.297	5.874	5.831	5.261	4.968	5.049	4.754	5.616	5.952
4.637	5.328	5.262	5.250	5.297	5.592	4.814	5.856	4.545	4.439	5.621	5.651	5.208	5.522
4.861	4.962	4.975	5.551	5.372	5.854	5.413	5.928	5.346	5.404	5.017	5.295	4.870	5.622
4.439	5.415	5.136	5.019	4.701	4.994	4.710	5.918	5.493	5.257	5.384	5.307	5.295	5.372
4.984	5.414	4.687	5.114	5.276	5.686	5.589	5.263	5.421	4.492	5.254	5.786	5.941	5.886
5.422	5.297	5.716	4.802	4.622	4.622	4.604	4.520	4.895	5.616	5.093	4.947	5.918	4.770
5.211	5.095	5.025	5.214	5.507	5.910	5.926	5.773	5.397	5.651	5.055	5.248	5.649	4.693
5.063	4.587	4.971	5.707	5.007	5.454	5.385	4.930	5.058	5.324	5.486	5.800	4.867	5.692
4.852	4.931	5.238	4.658	4.705	5.173	4.963	4.515	4.439	5.214	4.914	5.025	5.210	4.690
4.972	5.555	4.739	6.041	5.576	5.540	5.542	5.165	6.082	5.982	5.486	5.345	5.844	5.854
5.399	4.930	5.948	5.351	6.054	5.374	5.129	5.922	5.878	5.973	5.224	5.720	5.544	5.127
5.091													

(continued)

Table 4 (continued)

2006													
5.438	5.264	4.909	5.984	5.672	5.492	5.288	5.560	6.057	4.480	4.797	5.914	5.255	4.839
4.929	4.717	5.031	5.578	5.175	4.780	5.404	5.597	5.543	5.311	5.778	5.778	5.777	4.899
4.691	5.548	4.439	4.545	6.017	5.787	5.979	5.652	5.625	5.411	5.383	6.066	5.804	5.261
5.287	6.071	5.069	5.935	5.255	5.503	5.854	5.577	5.181	5.612	5.181	5.785	5.082	4.875
5.227	5.156	5.415	5.410	5.265	5.694	4.687	4.439	5.880	5.373	5.415	5.447	6.073	5.537
5.537	5.093	4.973	4.919	5.644	5.400	6.011	5.826	5.303	5.065	4.616	5.464	4.913	5.091
5.140	5.325	5.534	5.209	5.786	6.040	5.554	5.942	5.168	5.896	5.703	5.746	5.412	5.655
4.892	4.898	5.290	5.528	5.553	4.875	5.089	4.823	5.882	5.835	5.941	5.988	6.038	5.238
5.014	4.916	5.425	4.643	4.795	4.919	5.812	5.747	5.714	5.630	4.900	5.515	6.060	5.951
5.763	4.744	5.312	5.502	5.964	5.815	5.362	5.520	4.970	5.196	4.727	5.043	4.661	4.898
5.172	5.321	4.705	4.452	5.423	5.074	5.878	6.009	5.777	5.578	5.310	6.000	4.812	4.848
5.244	5.846	4.674	5.541	6.018	4.666	5.047	4.774	4.957	4.833	5.868	4.862	5.417	4.851
5.091	4.958	6.056	6.073	5.876	6.093	6.022	6.064	5.870	5.634	5.561	6.102	5.685	4.579
5.117	4.909	5.214	4.687	4.963	4.759	5.760	5.731	5.751	5.420	5.461	4.715	5.102	5.162
4.480	4.928	4.439	5.573	5.013	4.903	5.127	4.770	4.761	4.649	4.974	5.893	4.901	5.336
4.439	4.761	4.993	4.949	4.898	4.729	5.165	5.027	5.438	5.875	5.678	6.106	5.211	5.270
6.044	4.920	5.572	5.165	5.952	5.036	5.256	4.972	5.347	4.775	4.439	4.862	5.302	6.025
5.885	5.570	5.883	5.049	5.194	5.995	5.704	5.813	5.754	4.559	5.215	4.740	5.461	5.065
5.886	5.502	5.996	5.847	5.520	5.847	5.182	4.797	4.917	5.057	5.173	4.951	4.980	5.559
5.576	5.993	5.835	6.037	5.965	6.031	6.013	5.419	5.298	5.605	5.066	5.548	5.977	5.360
5.330	5.475	5.022	4.895	5.402	5.250	4.734	5.022	5.535	5.419	4.904	4.855	5.453	5.344
5.905	5.896	5.372	5.371	5.282	5.701	4.654	5.103	5.714	4.858	5.878	5.900	6.085	5.293
5.956	6.065	5.799	5.093	5.391	4.986	5.203	5.912	6.013	5.422	5.094	5.793	5.711	6.096
5.974	6.005	5.369	5.331	5.761	5.633	5.580	5.299	5.535	5.875	5.519	5.074	5.088	4.853
4.937	4.836	4.889	4.912	5.766	6.036	4.927	5.474	4.972	5.649	5.516	5.077	4.899	5.663
5.338	5.928	5.638	5.816	5.474	5.341	5.332	4.993	5.785	5.031	4.961	5.571	5.018	4.488
4.439													

(continued)

Table 4 (continued)

2007													
5.061	5.778	5.382	5.905	5.993	6.019	6.018	5.173	4.838	5.689	4.855	4.782	5.148	5.197
5.387	4.962	4.896	4.679	4.925	5.319	4.835	5.551	4.921	4.602	5.191	4.954	5.854	5.572
5.093	5.989	5.050	5.268	5.320	5.016	4.909	4.887	4.660	4.853	4.680	4.835	5.017	4.614
4.777	5.335	5.938	6.009	5.804	4.965	5.046	5.646	5.150	5.038	5.725	5.907	4.970	5.913
5.757	5.314	5.088	5.419	5.125	4.670	4.952	5.500	5.432	4.452	4.439	4.575	4.488	4.968
5.170	5.892	5.683	6.056	4.842	5.697	5.834	5.818	4.742	5.803	4.839	4.906	5.863	5.505
5.766	5.080	5.125	4.439	5.006	5.547	4.942	5.348	5.170	5.477	5.536	5.337	5.716	5.880
5.418	4.836	5.094	5.612	5.337	5.698	5.456	5.251	5.602	5.980	5.012	5.347	5.220	4.959
4.927	4.730	5.385	5.433	5.950	5.588	5.492	5.285	5.819	5.650	5.389	6.025	5.607	5.882
5.749	5.699	5.507	5.914	5.102	5.224	5.505	5.680	6.097	6.088	6.007	6.021	5.487	5.107
5.015	5.422	4.956	5.836	5.904	5.287	5.472	5.419	6.003	5.021	5.678	5.532	4.687	4.666
6.026	5.509	4.574	6.046	5.694	5.650	5.683	5.961	5.823	5.719	4.948	4.840	5.128	5.322
5.578	4.830	5.955	5.595	5.133	5.714	5.254	5.227	5.006	4.955	4.931	5.250	5.715	5.424
5.606	5.870	6.029	5.842	5.801	5.011	5.300	4.909	5.343	5.767	5.092	5.543	5.202	6.062
5.637	5.365	5.429	5.304	5.333	5.154	4.950	6.002	4.867	5.155	5.084	4.996	5.433	5.808
4.669	5.294	4.908	4.748	4.602	4.495	4.439	4.839	5.916	5.971	5.696	5.810	5.073	4.863
5.438	5.784	5.679	6.059	5.944	6.081	5.997	5.902	5.223	5.671	5.732	5.765	5.568	5.367
5.848	6.017	5.171	5.842	5.950	4.954	5.494	5.459	5.243	5.946	6.064	5.442	4.980	4.547
5.126	5.361	5.400	5.459	5.202	5.195	5.079	5.173	5.828	5.779	5.886	5.353	4.729	5.104
4.485	4.979	5.229	4.778	5.861	5.835	5.700	5.205	5.567	5.822	5.955	5.914	4.752	5.401
4.860	5.701	5.432	5.610	5.566	5.334	4.976	5.902	5.496	5.189	5.808	4.851	4.793	5.154
5.449	5.054	5.176	5.226	5.610	5.278	6.004	5.294	5.727	5.128	4.989	5.151	5.397	5.770
5.134	6.065	5.340	5.730	4.894	5.112	4.439	4.439	4.900	4.803	5.888	4.923	5.049	4.822
4.989	6.053	4.805	5.232	5.164	5.855	5.512	5.941	5.727	5.612	6.090	5.766	5.245	4.975
4.825	4.786	5.245	5.205	4.513	5.252	4.806	4.547	5.226	5.029	5.950	5.515	5.624	5.278
5.223	5.844	4.625	4.750	4.795	4.936	5.725	5.596	6.007	6.035	5.090	5.992	6.030	6.048
5.269													

(continued)

Table 4 (continued)

2008													
5.641	5.613	4.750	5.428	4.917	4.578	4.522	4.502	4.677	4.508	4.502	5.012	4.511	4.689
5.233	5.072	4.632	5.302	4.810	4.798	4.603	4.702	5.134	5.334	5.152	5.139	4.966	5.356
5.257	5.330	5.744	4.618	5.225	5.765	5.284	6.001	6.099	5.425	4.720	5.791	6.023	4.932
5.116	4.899	4.951	5.190	5.358	5.427	5.197	5.563	4.925	5.980	6.088	6.022	5.982	6.103
5.471	5.133	6.042	5.186	4.695	5.953	5.371	5.443	5.273	5.226	5.880	4.578	4.637	5.106
4.789	5.311	5.988	5.727	5.321	5.823	5.013	6.050	6.073	6.093	5.420	5.687	5.431	5.145
4.711	4.855	4.720	4.654	4.617	5.257	5.720	5.925	5.908	5.355	4.794	4.631	5.244	5.220
5.943	4.540	4.537	5.796	5.355	5.256	5.685	5.564	4.824	5.622	5.963	5.013	5.434	5.379
5.307	5.397	4.942	5.513	4.485	4.614	4.604	4.668	5.993	5.341	4.959	4.834	5.028	4.997
5.065	5.153	5.721	5.440	5.848	5.185	5.207	5.115	5.438	5.381	4.743	5.745	4.924	4.993
4.439	5.646	5.130	5.094	5.000	4.718	5.232	4.928	4.833	4.646	4.878	4.754	5.035	4.439
5.574	5.189	5.183	5.640	5.390	5.593	4.978	5.881	5.620	5.383	4.765	5.372	5.561	6.097
6.040	5.504	5.131	5.131	5.300	5.708	5.877	4.783	4.783	4.579	5.176	4.744	4.744	4.781
4.814	4.630	4.439	4.710	5.443	5.068	4.568	5.158	4.476	4.651	4.810	4.804	5.234	5.219
6.074	5.734	5.312	5.176	4.736	4.589	4.887	5.464	5.065	4.713	4.528	4.614	4.985	4.568
4.452	6.007	5.680	6.100	6.059	6.094	5.956	5.735	5.713	5.080	4.742	4.804	5.266	4.570
4.550	4.833	4.961	5.585	4.495	4.992	5.210	5.652	4.603	4.880	4.752	5.174	5.271	4.827
5.030	4.879	4.924	4.886	5.263	5.484	5.992	5.748	4.899	5.534	5.049	5.655	5.520	5.564
4.720	5.177	5.544	4.953	5.555	5.632	5.350	5.090	4.958	6.006	5.592	6.086	5.611	4.860
5.378	6.083	4.881	6.042	5.799	5.986	6.028	6.009	5.919	6.096	6.100	5.346	4.439	4.439
4.795	5.689	5.973	6.051	5.184	5.072	5.845	5.935	5.623	5.626	5.469	5.672	5.787	5.376
5.092	5.178	4.936	5.840	4.784	5.354	4.895	5.104	4.776	5.118	4.949	5.861	5.409	4.959
4.463	4.463	4.509	4.873	5.683	4.915	4.715	5.954	5.965	5.989	5.444	5.577	4.776	5.606
5.175	4.539	5.316	4.884	6.076	6.064	5.898	4.439	5.643	5.904	5.624	5.188	5.068	4.553
4.439	5.456	6.072	5.420	4.971	4.839	4.870	4.809	5.096	5.620	5.406	4.929	5.340	5.996
5.368	5.377	5.515	5.764	5.252	6.080	6.090	5.072	5.962	4.914	5.500	4.928	4.923	5.250
6.075	5.193												

(continued)

Table 4 (continued)

2009													
6.051	5.594	4.795	4.439	4.439	5.282	4.796	4.736	5.891	5.291	4.966	5.453	5.632	5.505
5.913	6.033	5.948	5.475	4.661	5.169	4.439	4.896	4.939	4.911	5.285	5.803	5.773	5.766
5.048	5.957	5.922	5.569	5.912	6.054	6.055	5.148	5.119	4.819	5.768	5.908	5.550	5.193
5.908	5.606	5.962	4.975	4.729	5.855	6.075	4.439	5.436	4.729	4.993	4.439	4.485	4.465
5.479	4.792	5.574	5.065	4.439	4.439	4.609	4.783	4.864	5.242	5.681	4.883	4.486	5.843
5.821	5.546	4.835	4.494	4.544	5.445	5.556	5.645	4.683	4.492	5.567	5.469	4.824	4.962
4.442	4.724	4.862	5.223	4.972	4.589	6.084	5.847	5.291	5.955	5.867	4.745	5.815	4.565
4.727	4.535	4.847	4.559	4.625	5.012	4.764	5.499	4.920	5.430	5.226	4.821	4.690	4.569
5.212	6.045	5.794	5.725	5.493	5.429	5.573	5.916	5.881	5.609	6.011	5.582	5.884	5.338
5.782	4.994	4.708	5.239	5.919	4.993	5.961	5.931	5.545	4.814	4.650	5.062	5.385	4.823
4.439	5.030	5.651	5.248	4.634	6.027	5.974	5.088	5.818	5.284	4.727	5.624	5.493	6.080
5.887	5.253	5.330	5.730	5.736	5.993	4.594	5.009	5.533	5.140	5.415	5.526	4.894	4.630
5.849	5.167	4.901	4.889	4.906	4.726	5.124	5.006	5.338	5.969	5.573	5.936	5.578	5.850
5.231	6.071	5.293	5.243	5.570	6.052	5.523	5.087	4.501	5.418	5.593	5.456	4.738	4.703
5.991	5.923	4.486	5.385	6.082	5.040	5.477	4.733	5.785	5.956	6.013	4.716	4.496	4.681
5.231	5.493	4.937	5.207	4.734	5.366	5.160	5.182	4.865	5.033	5.019	5.291	5.556	5.008
5.637	5.738	4.681	5.741	5.351	5.175	5.082	6.057	6.103	6.017	5.379	5.413	5.671	4.496
4.698	4.857	4.611	5.903	4.444	5.551	5.644	5.856	5.465	5.708	4.511	4.597	5.053	6.072
4.609	4.741	4.896	4.455	4.548	4.801	4.696	4.571	4.716	4.439	4.439	4.439	4.554	4.574
4.627	5.180	5.739	5.417	5.987	5.444	5.945	5.692	5.875	5.194	6.066	5.280	5.243	5.556
6.057	5.524	5.089	4.489	5.395	5.602	5.471	4.721	4.680	6.016	5.927	4.474	5.454	6.084
5.036	5.506	4.719	5.797	5.928	6.012	4.703	4.501	4.674	5.243	5.493	4.902	5.209	4.746
5.334	5.162	5.185	4.856	5.045	5.039	5.247	5.507	5.026	5.566	5.724	4.693	5.751	5.332
5.153	5.103	6.044	6.101	5.999	5.350	5.443	5.607	4.489	4.681	4.833	4.643	5.920	4.451
5.541	5.675	5.837	5.479	5.708	4.527	4.589	5.015	6.068	4.599	4.757	4.917	4.461	4.551
4.790	4.700	4.570	4.710	4.439	4.439	4.439	4.561	4.568	4.614	5.173	5.720	5.458	5.986
5.481													

(continued)

Table 4 (continued)

2010													
4.444	4.607	4.560	5.686	6.055	5.973	5.520	5.716	5.514	5.621	5.194	5.364	5.345	4.832
5.387	5.763	5.318	5.503	5.636	5.972	6.082	5.781	5.664	4.948	4.694	4.810	4.731	4.653
4.554	4.585	4.587	4.933	5.079	4.655	5.968	5.997	4.464	5.473	5.725	4.695	4.496	5.853
4.966	5.021	5.108	5.190	5.205	5.917	5.379	4.583	4.475	4.825	4.860	4.662	5.081	5.290
5.071	4.855	5.188	5.035	5.314	5.619	5.208	5.311	5.663	5.631	5.621	5.162	4.529	4.711
4.618	5.532	4.868	5.169	5.142	5.220	5.684	4.928	5.628	5.738	5.210	5.087	6.022	5.586
5.567	6.004	5.162	5.371	5.956	5.816	5.961	5.550	4.907	4.677	4.585	5.034	4.763	4.647
5.590	4.720	4.819	5.488	5.643	6.061	5.917	5.561	5.533	5.256	5.497	5.902	5.334	4.928
5.014	5.063	4.889	5.121	4.600	5.959	5.436	4.880	5.592	5.129	5.677	5.854	5.465	5.193
5.033	5.229	4.609	5.475	5.142	5.896	5.649	5.036	4.638	4.958	5.066	5.557	5.529	5.173
5.694	5.319	5.615	5.433	5.341	5.426	4.842	4.803	5.180	5.127	5.314	5.635	5.559	4.912
4.666	5.252	5.066	5.412	5.769	5.359	4.627	4.639	4.439	4.816	4.656	4.657	5.320	5.211
4.803	5.272	5.073	4.624	5.199	4.953	4.648	4.573	4.988	4.770	5.411	5.073	5.030	4.465
5.830	6.073	6.046	6.091	5.384	4.866	5.275	4.610	4.643	4.586	5.074	5.819	4.715	4.462
4.685	4.942	5.901	5.768	5.474	5.359	5.560	4.530	4.616	4.641	4.553	5.005	4.471	4.652
4.833	4.943	4.940	4.984	5.368	5.138	4.644	5.160	5.489	4.535	4.802	5.361	5.816	5.769
5.105	5.467	4.963	5.716	4.535	4.511	4.946	5.887	5.618	5.429	5.418	4.641	4.992	4.983
4.478	5.936	5.832	5.860	5.166	5.604	5.494	5.172	5.928	4.451	4.558	4.631	4.526	5.906
5.767	5.949	5.075	5.703	5.370	4.439	5.042	4.540	4.950	4.923	4.439	4.660	4.749	5.641
5.143	5.324	5.343	4.803	4.567	4.915	5.060	5.150	4.967	5.637	5.871	5.253	5.399	5.939
5.798	5.365	5.467	4.700	4.712	4.830	4.439	4.439	4.521	4.990	5.396	4.569	5.273	5.785
5.834	5.330	5.905	6.035	5.674	6.022	5.526	6.062	6.025	5.674	5.731	5.934	5.216	5.134
4.699	4.645	4.594	4.729	4.792	5.087	4.532	4.902	4.743	5.078	5.327	5.745	5.083	4.622
4.744	4.820	4.439	4.696	5.795	5.572	5.641	5.548	5.458	5.051	5.183	5.984	5.885	5.000
5.841	4.957	5.023	5.591	4.664	4.709	4.782	5.639	5.033	4.565	5.155	5.004	4.951	5.932
5.954	5.444	5.293	5.134	5.013	5.661	5.660	5.066	5.106	5.122	5.129	5.426	5.991	5.189
5.034													

Table 5 Entropy values prior crises

Day	26.02.01	27.02.01	28.02.01	01.03.01	02.03.01	03.03.01	04.03.01
Entropy	4.887	5.854	5.981	5.757	6.028	5.351	5.175
Day	16.09.08	17.09.08	18.09.08	19.09.08	20.09.08	21.09.08	22.09.08
Entropy	5.090	4.958	6.006	5.592	6.086	5.611	4.860

stock markets. The essence of the matter is as follows: the presence of the found pattern points to real possibility of big crisis beginning in few days.

Very appreciable book [5] states that all the crises at stock markets have many common prior markers. It is possible to say that the result of Sect. 5 is one of the mathematical expressions of this general assertion.

6 Conclusion

Let us give some remarks and comments to the above presented material.

1. It should be emphasized that the suggested approach has reasonably common character. This approach can be applied in various situations, in which system behavior can be presented in terms, expressed by the properties of a classification family, naturally determined by the functioning of the considered system. It is possible to suppose that the approach can be useful in attempts of sciences classification basing on analysis of big bulk of publications.

2. The applied results from Sects. 4 and 5 required different versions of the suggested general scheme. Another time it must be pointed to the necessity of manual parameters selection. Unfortunately, this is done by trials and errors method. It is supposed to improve it in future investigations.

3. It seems that additional use of some other parameters can get rid of very few superfluous noncrisis days with crisis pattern. It will be done in future investigations.

4. Some elements of the general scheme can be improved. Partially, in formulae (5) and (7) it is possible to define value of parameter s in dependence of convergence of the corresponding index values. This version can significantly reduce calculation time.

5. All the experimental results concern only one parliament and only one stock market. Of course, in the further investigation it is supposed to consider essentially wider data sets. It will be done as the required information becomes available.

6. The last remark is as follows. Several times in the article the expression "big crisis" was mentioned without any formal definition. It seems that to give a formal definition of this notion is practically impossible. However, such a definition is unessential. We can simply suppose that a big crisis is a ***state of the stock market, which the most of participants perceive as a big crisis***. Just this assumption leads to their behavior, characterized by the found pattern.

Acknowledgements The author is grateful to F.T. Aleskerov for his support and attention to the work, V.I. Jakuba for help in program realizations, G.I. Penikas for helpful discussions, and E. Misharina for S&P-500 data extraction. The article was prepared within the framework of the Basic Research Program at the National Research University Higher School of Economics (HSE) and supported within the framework of a subsidy granted to the HSE by the Government of the Russian Federation for the implementation of the Global Competitiveness Program.

References

1. Feller, W.: An Introduction to Probability Theory and Its Applications. Revised Printing. Wiley (1970)
2. Luxburg, U.: A tutorial on spectral clustering. Stat. Comput. **17**(4), 395–416 (2007)
3. Mirkin, B.: Clustering: A Data Recovery Approach. Chapman & Hall/CRC (2012)
4. Mirkin, B.: Core Concepts in Data Analysis: Summarization, Correlation. Visualization. Springer (2011)
5. Reinhart, C., Rogoff, K.: This Time is Different. Princeton University Press (2009)
6. Rubchinsky, A.: Divisive-Agglomerative Classification Algorithm Based on the Minimax Modification of Frequency Approach: Preprint/NRU HSE/WP7/2010/07. M.: 2010. 48 p. //Available in Internet at the address: www.hse.ru/en/org/hse/wp/wp7en
7. Rubchinsky, A.: Divisive-Agglomerative Algorithm and Complexity of Automatic Classification Problems: Preprint /NRU HSE/WP7/2015/09. M.: 2015. 44 p. Available in Internet at the address: www.hse.ru/en/org/hse/wp/wp7en

Homogeneity Hypothesis Testing for Degree Distribution in the Market Graph

D.P. Semenov and P.A. Koldanov

Abstract The problem of homogeneity hypothesis testing for degree distribution in the market graph is studied. Multiple hypotheses testing procedure is proposed and applied for China and India stock markets. The procedure is constructed using bootstrap method for individual hypotheses and Bonferroni correction for multiple testing. It is shown that homogeneity hypothesis of degree distribution for the stock markets for the period of 2003–2014 is not accepted.

Keywords Homogeneity hypothesis · Degree distribution · Market graph · Bootstrap method · Multiple hypotheses testing · Bonferroni correction

1 Introduction

Market graph was proposed in [2] as a tool for stock market analysis. The dynamic of the market graph for US stock market was studied in [4], where edges density, maximum cliques, and maximum independent sets of the market graph were considered. There are another characteristics of the market graph which are interesting in market network analysis. In this paper, we investigate the degree distribution of vertices in the market graph. From economic point of view, the degree of vertex characterizes the influence of the corresponding stock on the stock market. For example, the network topology structure as a star means the presence of a dominating stock. On the other hand, uniform distribution of degrees of vertices can be interpreted as a characteristic of "free" market.

In this paper, we investigate the problem of stationarity of network topologies over time. The main question is: are there statistically significant differences in the topology of the market graphs for different periods of observation? The problem

D.P. Semenov (✉) · P.A. Koldanov
Laboratory of Algorithms and Technologies for Network Analysis,
National Research University Higher School of Economics, Nizhny Novgorod, Russia
e-mail: dimsem2010@yandex.ru

P.A. Koldanov
e-mail: pkoldanov@hse.ru

V.A. Kalyagin et al. (eds.), *Models, Algorithms, and Technologies for Network Analysis*, Springer Proceedings in Mathematics & Statistics 197,
DOI 10.1007/978-3-319-56829-4_11

of homogeneity of degree distributions over time is considered as multiple testing problems of homogeneity hypotheses of degree distributions for each pairs of years. At the same time, the problem of homogeneity hypotheses of degree distributions for pair of years is considered as multiple testing problem of homogeneity hypotheses for each vertex degree.

To test the homogeneity hypotheses for each vertex degree the method based on the confidence intervals is applied. To construct confidence intervals the bootstrap method is used. In order to construct a multiple testing procedure with a given significance level we use Bonferroni corrections. The obtained procedure is applied to China and India stock markets for the period from 2003 to 2014 (12 years). To conduct experiments 100 most liquid stocks are selected from each market.

The paper is organized as follows. In Sect. 2 a brief overview of the market graph approach is given. In Sect. 3, we formally state the problem. In Sect. 4 a detailed descriptions of the multiple testing statistical procedure for testing homogeneity hypotheses of degree distribution is given. In Sect. 5, the results of application of this procedure to Chinese and Indian stock market are presented. The Sect. 6 summarizes the main results of the paper.

2 Market Graph Model

Let N be the number of stocks on the stock market. Let $p_i(t)$ be the price of the stock i for the day t, and $r_i(t)$ be the log return of the stock i for the day t:

$$r_i(t) = \log \frac{p_i(t)}{p_i(t-1)}$$

We assume that $r_i(t)$ are observations of the random variables $R_i(t)$, random variables $R_i(t)$, $t = 1, 2, \ldots, n$ are independent and identically distributed as R_i for fixed i, and random vector $(R_1, R_2, \ldots, R_N)$ has a multivariate normal distribution with correlation matrix $||\rho_{i,j}||$.

Let

$$r_{i,j} = \frac{\Sigma(r_i(t) - \overline{r_i})(r_j(t) - \overline{r_j})}{\sqrt{\Sigma(r_i(t) - \overline{r_i})^2}\sqrt{\Sigma(r_j(t) - \overline{r_j})^2}}$$

be the estimated value of correlation coefficient between returns of the stocks i and j, where

$$\overline{r_i} = \frac{1}{n}\sum_{t=0}^{n} r_i(t)$$

Matrix $||\rho_{i,j}||$ is used to construct a true market graph, while matrix $||r_{i,j}||$ is used to construct a sample market graph. The procedure of the market graph construction is the following. Each vertex represents a stock. An edge connects two vertices i and j, if $||\rho_{i,j}|| > \rho_0$ in case of the true market graph, and if $||r_{i,j}|| > r_0$ (where ρ_0, r_0 are threshold values) in case of the sample market graph. When the vertices share a common edge, they are called adjacent.

3 Problem Statement

For a market graph on N vertices one can associate the following two-dimensional array:

$$\begin{array}{cccc} 0 & 1 & \dots & N-1 \\ v_0 & v_1 & \dots & v_{N-1} \end{array} \tag{1}$$

where line 1 represents the degree of vertices and the line 2 represents the number of vertices of the given degree. Denote by F_v the vector of degree distribution of vertices, $F_v = (v_0, v_1, \dots, v_{N-1})$.

Let L be the number of different periods of observations. The hypothesis of homogeneity of degree distributions over L periods of observations can be written as:

$$H_0 : F_v^1 = F_v^2 = \dots = F_v^L \tag{2}$$

where F_v^l is the distribution of vertex degrees for the period of observation$l$, $l = 1, 2, \dots, L$.

The problem of testing H_0 could be considered as multiple testing problem for individual homogeneity hypotheses:

$$h^{k,l} : F_v^k = F_v^l, \;\; k, l = 1, 2, \dots, L, \;\; k \neq l \tag{3}$$

The hypothesis $h^{k,l}$ is the homogeneity hypothesis of degree distributions for the pair of years k and l. Hypothesis H_0 can be presented as the intersection of hypotheses $h^{k,l}$:

$$H_0 = \bigcap_{k,l=1,2,\dots,L,k\neq l} h^{k,l}$$

In this case, hypothesis H_0 is accepted if and only if all individual hypotheses $h^{k,l}$ are accepted, and hypothesis H_0 is rejected if at least one individual hypothesis $h^{k,l}$ is rejected.

The problem of testing individual hypothesis $h^{k,l}$ (homogeneity hypotheses of degree distributions for the pair of years k and l) can be considered as multiple testing problem of individual homogeneity hypotheses for each vertex degree:

$$h_j^{k,l} : v_j^k = v_j^l, \quad j = 0, 1, 2, \dots, N-1$$

One can consider the hypothesis $h^{k,l}$ as the intersection of individual hypotheses $h_j^{k,l}$

$$h^{k,l} = h_0^{k,l} \cap h_1^{k,l} \ldots \cap h_{N-1}^{k,l}$$

In this case, hypothesis $h^{k,l}$ is accepted if and only if all individual hypotheses $h_j^{k,l}$, $j = 0, 1, 2, \ldots, N-1$ are accepted, and hypothesis $h^{k,l}$ is rejected if at least one individual hypothesis $h_j^{k,l}$ is rejected.

4 Statistical Procedure for Homogeneity Hypotheses Testing

Consider, the individual hypotheses of the following form:

$$h^{k,l} : F_v^k = F_v^l \quad (4)$$

Let R^k, R^l be random vectors of distributions of stock returns for the periods k and l respectively. In order to test (4) we use two sequences of n_1 and n_2 observations of random vectors R^k and R^l (in what follows we suppose for simplicity $n_1 = n_2 = n$):

$$\begin{pmatrix} r_1^k(1) \\ r_2^k(1) \\ \ldots \\ r_N^k(1) \end{pmatrix} \begin{pmatrix} r_1^k(2) \\ r_2^k(2) \\ \ldots \\ r_N^k(2) \end{pmatrix} \ldots \begin{pmatrix} r_1^k(n) \\ r_2^k(n) \\ \ldots \\ r_N^k(n) \end{pmatrix} \quad (5)$$

$$\begin{pmatrix} r_1^l(1) \\ r_2^l(1) \\ \ldots \\ r_N^l(1) \end{pmatrix} \begin{pmatrix} r_1^l(2) \\ r_2^l(2) \\ \ldots \\ r_N^l(2) \end{pmatrix} \ldots \begin{pmatrix} r_1^l(n) \\ r_2^l(n) \\ \ldots \\ r_N^l(n) \end{pmatrix} \quad (6)$$

where $r_i^k(t)$ is the return of the stock i in the day t for the year k and $r_i^l(t)$ is the return of the stock i in the day t for the year l.

Using these observations, we construct the sample market graphs with a given threshold for the periods k and l and calculate its degree distributions. We use these sample degree distributions to construct individual test for hypothesis $h_j^{k,l}$. The individual test for $h_j^{k,l}$ will use a confidence intervals for ν_j^k and ν_j^l. To construct these confidence intervals we apply bootstrap procedure [5] in the following way:

1. Apply S times the statistical bootstrap procedure for each sequence of observations (5) and (6).
2. For each bootstrap sample, calculate the sample market graph and find the number of vertices of degree j in the sample market graph.
3. Calculate α-confidence interval for the number of vertices with degree j.

To take the decision for the hypothesis $h_j^{k,l}$, we use the following procedure: if the confidence intervals for ν_j^k and ν_j^l do not intersect, then the hypothesis is rejected.

Otherwise it is accepted. Individual hypothesis $h^{k,l}$ is accepted if all hypotheses $h_j^{k,l}$, $j = 0, 1, 2, \ldots, N-1$ are accepted. Finally, the hypothesis H_0 is accepted if all hypotheses $h^{k,l}$ are accepted.

Let us introduce some notations. Define indicator of vertex degree in a sample graph as follows ($i = 1, 2, \ldots, N, j = 0, 1, \ldots, N-1$):

$$\chi_{i,j} = \begin{cases} 1, \text{ if vertex } i \text{ has degree } j \\ 0, \text{ otherwise} \end{cases} \tag{7}$$

Distribution of vertex degrees in one of bootstrap samples q ($q = 1, 2, \ldots, S$) for the period of observation k is defined by:

$$\nu_0^k(q), \nu_1^k(q), \ldots, \nu_{N-1}^k(q)$$

with

$$\nu_j^k(q) = \sum_{i=1}^{N} \chi_{i,j}^k(q)$$

Using asymptotic normal approximation one can write the test for the hypothesis $h_j^{k,l}$ in the following form

$$\varphi_j^{k,l} = \begin{cases} 0, \text{ if } |\overline{\nu}_j^k - \overline{\nu}_j^l| < c(\alpha')(\sigma(\nu_j^k) + \sigma(\nu_j^l)) \\ 1, \text{ otherwise} \end{cases} \tag{8}$$

where

$$\overline{\nu}_j^k = \frac{1}{S}\sum_q \nu_j^k(q), \quad \overline{\nu}_j^l = \frac{1}{S}\sum_q \nu_j^l(q)$$

and $c(\alpha')$ is $(1-\alpha')$-two size quantile of standard normal distribution. For example, for $\alpha' = 0, 05$ and $c_{\alpha'} = 0, 98$.

When we deal with hypothesis $h^{k,l}$ we face with the multiple testing problem of homogeneity hypotheses for each vertex degree. To control the probability of first type error Bonferroni correction is used. This means that significance level α' for hypothesis $h_j^{k,l}$ is chosen as follows $\alpha' = \alpha/100$, where α is the significance level of the resulting test for the hypothesis $h^{k,l}$. To test the hypothesis H_0 with the probability of the first type error α one has to choose the error rate for the tests $\varphi_j^{k,l}$ equal to $\alpha'' = \alpha/(100 * C_L^2)$ (double Bonferroni correction).

5 Experimental Results

The experiments are conducted on the basis of data from the stock markets of China and India. The 100 most traded stocks for the period from 01 January 2003 to 31 December 2014 are considered. The number of observed days $n = 250$ (1 calendar year). The results are shown in the tables below. In each table element (k, l) is equal to zero if the hypothesis $h^{k,l}$ is accepted and equal to 1 otherwise. Tables 1, 2, 3, 4, 5 and 6 present the results for a different values of threshold for Chinese stock market. Tables 7, 8, 9, 10, 11 and 12 present the results for a different values of threshold for Indian stock market. One can see that pairwise hypotheses of homogeneity are mostly rejected. If the value of threshold is increasing then more and more homogeneity hypotheses are accepted.

Pairwise hypotheses of homogeneity mainly rejected. However, there are 2 years (2003, 2007), for which the homogeneity hypotheses are accepted for selected values of threshold. For (2003, 2013) there are thresholds, for which the homogeneity hypotheses are accepted and are rejected.

Table 1 Threshold=0.2, Chinese market. 0—acceptance of hypothesis, 1—rejection of hypothesis

	2003	2004	2005	2006	2007	2008	2009	2010	2011	2012	2013	2014
2003	0	1	0	1	0	1	1	1	1	1	0	1
2004	1	0	0	1	1	1	1	1	1	1	1	1
2005	0	0	0	1	1	1	1	1	1	1	1	1
2006	1	1	1	0	1	1	1	1	0	1	1	1
2007	0	1	1	1	0	1	1	1	1	1	1	1
2008	1	1	1	1	1	0	1	0	0	1	1	1
2009	1	1	1	1	1	1	0	1	1	1	1	1
2010	1	1	1	1	1	0	1	0	0	1	1	1
2011	1	1	1	0	1	0	1	0	0	1	1	1
2012	1	1	1	1	1	1	1	1	1	0	0	1
2013	0	1	1	1	1	1	1	1	1	0	0	1
2014	1	1	1	1	1	1	1	1	1	1	1	0

Table 2 Threshold=0.3, Chinese market. 0—acceptance of hypothesis, 1—rejection of hypothesis

	2003	2004	2005	2006	2007	2008	2009	2010	2011	2012	2013	2014
2003	0	1	0	1	0	1	1	1	1	1	0	1
2004	1	0	0	1	1	1	1	1	1	1	1	1
2005	0	0	0	1	1	1	1	1	1	1	1	1
2006	1	1	1	0	1	1	1	1	0	1	1	1
2007	0	1	1	1	0	1	1	0	1	1	1	1
2008	1	1	1	1	1	0	1	0	0	1	1	1
2009	1	1	1	1	1	1	0	1	1	1	1	1
2010	1	1	1	1	0	0	1	0	0	1	1	1
2011	1	1	1	0	1	0	1	0	0	1	1	1
2012	1	1	1	1	1	1	1	1	1	0	0	1
2013	0	1	1	1	1	1	1	1	1	0	0	1
2014	1	1	1	1	1	1	1	1	1	1	1	0

Table 3 Threshold=0.4, Chinese market. 0—acceptance of hypothesis, 1—rejection of hypothesis

	2003	2004	2005	2006	2007	2008	2009	2010	2011	2012	2013	2014
2003	0	1	1	1	0	1	1	1	1	1	1	1
2004	1	0	0	1	1	1	1	0	1	1	1	0
2005	1	0	0	1	0	1	0	0	1	1	1	0
2006	1	1	1	0	1	1	1	1	1	1	1	1
2007	0	1	0	1	0	1	1	1	1	1	0	0
2008	1	1	1	1	1	0	1	1	1	1	1	1
2009	1	1	0	1	1	1	0	1	1	1	1	1
2010	1	0	0	1	1	1	1	0	1	1	1	1
2011	1	1	1	1	1	1	1	1	0	1	1	1
2012	1	1	1	1	1	1	1	1	1	0	1	1
2013	1	1	1	1	0	1	1	1	1	1	0	1
2014	1	0	0	1	0	1	1	1	1	1	1	0

Table 4 Threshold=0.5, Chinese market. 0—acceptance of hypothesis, 1—rejection of hypothesis

	2003	2004	2005	2006	2007	2008	2009	2010	2011	2012	2013	2014
2003	0	1	1	1	0	1	0	1	1	1	0	0
2004	1	0	0	1	1	1	0	1	1	1	1	1
2005	1	0	0	1	1	1	1	0	1	1	1	1
2006	1	1	1	0	1	1	1	1	1	1	1	1
2007	0	1	1	1	0	1	1	1	1	1	1	0
2008	1	1	1	1	1	0	1	1	1	1	1	1
2009	0	0	1	1	1	1	0	1	1	1	0	0
2010	1	1	0	1	1	1	1	0	1	1	1	0
2011	1	1	1	1	1	1	1	1	0	1	1	1
2012	1	1	1	1	1	1	1	1	1	0	1	1
2013	0	1	1	1	1	1	0	1	1	1	0	1
2014	0	1	1	1	0	1	0	0	1	1	1	0

Table 5 Threshold=0.6, Chinese market. 0—acceptance of hypothesis, 1—rejection of hypothesis

	2003	2004	2005	2006	2007	2008	2009	2010	2011	2012	2013	2014
2003	0	1	1	1	0	1	0	1	1	1	0	0
2004	1	0	0	1	1	1	0	1	1	1	1	1
2005	1	0	0	1	1	1	1	0	1	1	1	1
2006	1	1	1	0	1	1	1	1	1	1	1	1
2007	0	1	1	1	0	1	1	1	1	1	1	0
2008	1	1	1	1	1	0	1	1	1	1	1	1
2009	0	0	1	1	1	1	0	1	1	1	0	0
2010	1	1	0	1	1	1	1	0	1	1	1	0
2011	1	1	1	1	1	1	1	1	0	1	1	1
2012	1	1	1	1	1	1	1	1	1	0	1	1
2013	0	1	1	1	1	1	0	1	1	1	0	1
2014	0	1	1	1	0	1	0	0	1	1	1	0

Table 6 Threshold=0.7, Chinese market. 0—acceptance of hypothesis, 1—rejection of hypothesis

	2003	2004	2005	2006	2007	2008	2009	2010	2011	2012	2013	2014
2003	0	0	1	1	0	1	1	1	0	1	1	1
2004	0	0	0	0	0	1	1	1	0	1	1	1
2005	1	0	0	0	0	1	1	1	0	1	1	1
2006	1	0	0	0	0	1	1	1	0	1	1	1
2007	0	0	0	0	0	1	1	1	0	1	1	1
2008	1	1	1	1	1	0	1	1	1	1	1	1
2009	1	1	1	1	1	1	0	0	1	1	0	0
2010	1	1	1	1	1	1	0	0	1	1	0	0
2011	0	0	0	0	0	1	1	1	0	1	1	1
2012	1	1	1	1	1	1	1	1	1	0	0	0
2013	1	1	1	1	1	1	0	0	1	0	0	0
2014	1	1	1	1	1	1	0	0	1	0	0	0

Table 7 Threshold=0.2, Indian market. 0—acceptance of hypothesis, 1—rejection of hypothesis

	2003	2004	2005	2006	2007	2008	2009	2010	2011	2012	2013	2014
2003	0	1	1	1	1	1	1	1	1	1	1	1
2004	1	0	1	1	1	1	1	1	1	1	1	1
2005	1	1	0	1	1	1	0	1	1	1	1	1
2006	1	1	1	0	1	1	0	1	1	1	1	1
2007	1	1	1	1	0	1	1	1	1	1	1	1
2008	1	1	1	1	1	0	1	1	1	1	1	1
2009	1	1	0	0	1	1	0	1	0	0	0	1
2010	1	1	1	1	1	1	1	0	1	1	1	1
2011	1	1	1	1	1	1	0	1	0	0	1	1
2012	1	1	1	1	1	1	0	1	0	0	0	1
2013	1	1	1	1	1	1	0	1	1	0	0	1
2014	1	1	1	1	1	1	1	1	1	1	1	0

Table 8 Threshold=0.3, Indian market. 0—acceptance of hypothesis, 1—rejection of hypothesis

	2003	2004	2005	2006	2007	2008	2009	2010	2011	2012	2013	2014
2003	0	1	1	1	1	1	1	1	1	1	1	1
2004	1	0	1	1	1	1	1	1	1	1	1	1
2005	1	1	0	1	1	1	1	1	1	1	1	1
2006	1	1	1	0	1	1	1	1	1	1	1	1
2007	1	1	1	1	0	1	1	1	1	1	1	1
2008	1	1	1	1	1	0	1	1	1	1	1	1
2009	1	1	1	1	1	1	0	1	1	1	0	1
2010	1	1	1	1	1	1	1	0	1	1	1	1
2011	1	1	1	1	1	1	1	1	0	0	1	0
2012	1	1	1	1	1	1	1	1	0	0	0	1
2013	1	1	1	1	1	1	0	1	1	0	0	0
2014	1	1	1	1	1	1	1	1	0	1	0	0

Table 9 Threshold=0.4, Indian market. 0—acceptance of hypothesis, 1—rejection of hypothesis

	2003	2004	2005	2006	2007	2008	2009	2010	2011	2012	2013	2014
2003	0	1	1	1	1	1	1	1	1	1	1	1
2004	1	0	1	1	1	1	1	1	1	1	0	1
2005	1	1	0	1	1	1	1	1	1	1	1	1
2006	1	1	1	0	1	1	1	1	1	1	1	1
2007	1	1	1	1	0	1	1	1	1	1	1	1
2008	1	1	1	1	1	0	1	1	1	1	1	1
2009	1	1	1	1	1	1	0	1	1	1	0	1
2010	1	1	1	1	1	1	1	0	1	1	1	1
2011	1	1	1	1	1	1	1	1	0	0	0	0
2012	1	1	1	1	1	1	1	1	0	0	0	0
2013	1	0	1	1	1	1	0	1	0	0	0	0
2014	1	1	1	1	1	1	1	1	0	0	0	0

Table 10 Threshold=0.5, Indian market. 0—acceptance of hypothesis, 1—rejection of hypothesis

	2003	2004	2005	2006	2007	2008	2009	2010	2011	2012	2013	2014
2003	0	1	0	1	1	1	1	1	1	1	1	1
2004	1	0	1	1	1	1	0	1	1	1	1	0
2005	0	1	0	1	1	1	1	0	1	1	1	1
2006	1	1	1	0	1	1	1	1	1	1	1	1
2007	1	1	1	1	0	1	1	0	1	1	1	1
2008	1	1	1	1	1	0	1	1	1	1	1	1
2009	1	0	1	1	1	1	0	1	0	1	1	1
2010	1	1	0	1	0	1	1	0	1	1	1	1
2011	1	1	1	1	1	1	0	1	0	1	0	0
2012	1	1	1	1	1	1	1	1	1	0	0	0
2013	1	1	1	1	1	1	1	1	0	0	0	0
2014	1	0	1	1	1	1	1	1	0	0	0	0

Table 11 Threshold=0.6, Indian market. 0—acceptance of hypothesis, 1—rejection of hypothesis

	2003	2004	2005	2006	2007	2008	2009	2010	2011	2012	2013	2014
2003	0	1	0	1	1	1	1	1	1	1	1	1
2004	1	0	1	1	1	1	0	1	1	1	1	0
2005	0	1	0	1	1	1	1	0	1	1	1	1
2006	1	1	1	0	1	1	1	1	1	1	1	1
2007	1	1	1	1	0	1	1	0	1	1	1	1
2008	1	1	1	1	1	0	1	1	1	1	1	1
2009	1	0	1	1	1	1	0	1	0	1	1	1
2010	1	1	0	1	0	1	1	0	1	1	1	1
2011	1	1	1	1	1	1	0	1	0	1	0	0
2012	1	1	1	1	1	1	1	1	1	0	0	0
2013	1	1	1	1	1	1	1	1	0	0	0	0
2014	1	0	1	1	1	1	1	1	0	0	0	0

Table 12 Threshold=0.7, Indian market. 0—acceptance of hypothesis, 1—rejection of hypothesis

	2003	2004	2005	2006	2007	2008	2009	2010	2011	2012	2013	2014
2003	0	1	0	0	0	1	1	0	0	1	1	1
2004	1	0	1	0	1	0	0	1	0	0	1	0
2005	0	1	0	1	0	1	1	0	0	1	1	1
2006	0	0	1	0	1	0	0	1	0	0	0	0
2007	0	1	0	1	0	1	1	0	0	1	1	1
2008	1	0	1	0	1	0	0	1	1	0	1	1
2009	1	0	1	0	1	0	0	1	0	0	1	1
2010	0	1	0	1	0	1	1	0	0	1	1	1
2011	0	0	0	0	0	1	0	0	0	0	1	1
2012	1	0	1	0	1	0	0	1	0	0	0	0
2013	1	1	1	0	1	1	1	1	1	0	0	0
2014	1	0	1	0	1	1	1	1	1	0	0	0

6 Conclusions

In this paper, we investigated the homogeneity of degree distribution in the market graph over time. The procedure of comparison of degree distributions for different periods of observation was built to study this problem. This procedure has been applied to the real data yields the 100 most traded shares for Chinese and Indian stock markets. Conducted experiments show that vertex degree distribution is not stationary and significantly changes over the time.

Acknowledgements The work of Koldanov P.A. was conducted at the Laboratory of Algorithms and Technologies for Network Analysis of National Research University Higher School of Economics. The work is partially supported by RFHR grant 15-32-01052.

References

1. Anderson, T.W.: An Introducion to Multivariate Statistical Analysis, 3rd edn. Wiley-Interscience, New York (2003)
2. Boginsky, V., Butenko, S., Pardalos, P.M.: On structural properties of the market graph. In: Nagurney, A. (ed.) Innovations in Financial and Economic Networks, pp. 29–45. Edward Elgar Publishing Inc., Northampton (2003)
3. Boginsky, V., Butenko, S., Pardalos, P.M.: Statistical analysis of financial networks. Comput. Stat. Data Anal. **48**, 431–443 (2005)
4. Boginsky, V., Butenko, S., Pardalos, P.M.: Mining market data: a network approach. Comput. Oper. Res. **33**, 3171–3184 (2006)
5. Efron, B.: Bootstrap methods: another look at the Jackknife. Ann. Stat. **7**(1), 126 (1979)

Stability Testing of Stock Returns Connections

M.A. Voronina and P.A. Koldanov

Abstract The problem of stability of connections of stock returns over time is considered. This problem is formulated as a multiple testing problem of homogeneity of covariance matrices. A statistical procedure based on Box's M-test and Bonferroni correction is proposed. This procedure is applied to French and German stock markets.

1 Introduction

The study of economic networks and understanding of its structural properties are essential for financial and investment decision making, construction of an optimal investment portfolio and extraction of other important information about a market. Stock market network model is a complete weighted graph, graph nodes correspond to the stocks, and weights of edges are estimated as covariances of stock returns [2].

In this paper, stock returns are random variables and their joint distribution is multivariate normal distribution with mean vector μ, and covariance matrix Σ. It is assumed that there are several network stock market observations in different time intervals. Our main problem is to test the homogeneity hypothesis over time for covariance matrices of stock returns. To handle the problem we suggest to use Box's M-test for testing homogeneity of two covariance matrices and the Bonferroni correction to combine individual tests in multiple testing procedure. The obtained procedure controls *FWER (Family Wise Error Rate)* at a given significance level α. This procedure is applied to the French and German stock markets with observations over the period 2003–2014. Numerical results provide support to the nonhomogeneity

M.A. Voronina (✉) · P.A. Koldanov
National Research University Higher School of Economics, Laboratory of Algorithms and Technologies for Network Analysis, Nizhny Novgorod, Moscow, Russia
e-mail: voromari@yandex.ru

P.A. Koldanov
e-mail: pkoldanov@hse.ru

V.A. Kalyagin et al. (eds.), *Models, Algorithms, and Technologies for Network Analysis*, Springer Proceedings in Mathematics & Statistics 197,
DOI 10.1007/978-3-319-56829-4_12

hypothesis over the overall period 2003–2014, but it is shown that there are special groups of companies and periods, where the homogeneity hypothesis is confirmed.

The paper is organized as following. In Sect. 2 main notations and problem statement are given. In Sect. 3 Box's M-test is described. In Sect. 4 some properties of Box's M-test are investigated. In Sect. 5 the results of application of the obtained multiple testing procedure to the French and German stock markets are given. In Sect. 6 some concluding remarks are presented.

2 Problem Statement

Suppose, we have q time intervals of observations and multivariate distribution of stock returns is $N(\mu^{(g)}, \Sigma_g)$ for the time interval g, $g = 1, 2, \ldots, q$. Let $x_k^{(g)} (k = 1, \ldots, N_g; g = 1, \ldots, q)$ be a sample of the size N_g from $N(\mu^{(g)}, \Sigma_g)$. The main goal of the article is to test the homogeneity hypothesis

$$H_0 : \Sigma_1 = \cdots = \Sigma_q \tag{1}$$

where Σ_g is the covariance matrix of stock returns for the time period $g, g = 1, \ldots, q$

We consider the problem (1) as a multiple testing problem of the pairwise homogeneity hypotheses. We use the probability of at least one false rejection *FWER(family-wise error rate)* to control error of a multiple testing procedure. We formulate individual hypothesis as:

$$h_{i,j} : \Sigma_i = \Sigma_j \ vs \ k_{i,j} : \Sigma_i \neq \Sigma_j; i, j = 1, \ldots, q \tag{2}$$

Hypothesis H_0 is accepted if and only if all individual hypotheses (2) are accepted. To test individual hypothesis, we use the Box's M-test [3] with a given significance level α'. To control the FWER of the multiple testing procedure we use the Bonferroni correction $\alpha' = \frac{\alpha}{m}$; $m = C_q^2$. In this case FWER of the multiple testing procedure is bounded above by α [4].

3 Box's M-Test for Two Covariance Matrices

Let

$x_k^{(1)} (k = 1, \ldots, N_1)$ be a sample from multivariate normal distribution $N(\mu^{(1)}, \Sigma_1)$, and

$x_k^{(2)} (k = 1, \ldots, N_2)$ be a sample from multivariate normal distribution $N(\mu^{(2)}, \Sigma_2)$, where μ are unknown. In this case null hypothesis H_0 takes the form

$$H_0 : \Sigma_1 = \Sigma_2 \tag{3}$$

To construct the Box's M-test, we use the following M statistic [1]:

$$M = (N_1 + N_2)ln\Big(|\frac{A_1 + A_2}{N_1 + N_2}|\Big) - \Big((N_1 - 1)ln\Big(|\frac{1}{N_1 - 1}A_1|\Big) + (N_2 - 1)ln\Big(|\frac{1}{N_2 - 1}A_2|\Big)\Big),$$

where

$$A_1 = \sum_{\alpha=1}^{N_1}\Big(x_\alpha^{(1)} - \overline{x^{(1)}}\Big)\Big(x_\alpha^{(1)} - \overline{x^{(1)}}\Big)', \quad A_2 = \sum_{\alpha=1}^{N_2}\Big(x_\alpha^{(2)} - \overline{x^{(2)}}\Big)\Big(x_\alpha^{(2)} - \overline{x^{(2)}}\Big)'.$$

Let us introduce some notations:

$$c = \frac{2p^2 + 3p - 1}{6(p+1)}\Big(\frac{1}{N_1 - 1} + \frac{1}{N_2 - 1} - \frac{1}{N_1 + N_2 - 2}\Big),$$

$$c_2 = \frac{(p-1)(p+2)}{6}\Big(\frac{1}{(N_1 - 1)^2} + \frac{1}{(N_2 - 1)^2} - \frac{1}{(N_1 + N_2 - 2)^2}\Big),$$

$$df = \frac{p(p+1)}{2}, \; df_2 = \frac{df + 2}{c_2 - c^2},$$

$$a^+ = \frac{df}{1 - c - \frac{df}{df_2}}, \; F^+ = \frac{M}{a^+},$$

$$a^- = \frac{df_2}{1 - c + \frac{2}{df_2}}, \; F^- = \frac{df_2 \cdot M}{df(a^- - M)},$$

$$F = \begin{cases} F^+, \text{ if } c_2 > c^2; \\ F^-, \text{ if } c_2 < c^2. \end{cases}$$

Asymptotic distribution of the statistic F is known to be Fisher distribution with (df, df_2) degrees of freedom [3]. Therefore the null hypothesis is rejected, if $F > F(df, df_2)(\alpha)$, where $F(df, df_2)(\alpha)$ is $(1 - \alpha)$-quantile of the Fisher distribution. This is the Box's M-test for two covariance matrices.

4 Investigation of the Probability of Type I Error for Box's M-Test

Box's M-test uses asymptotic distribution of the statistic F. It is possible that the test does not control the given significance level for a large number of stocks (dimension of covariance matrices) and a small sample sizes. In this section, we are interested in the probability of false rejection (Type I error) of the hypothesis H_0 as a function of the

dimension of covariance matrix and of the sample size (we suppose that $N_1 = N_2$). For estimation of the probability of Type I error we use statistical simulations and apply the following algorithm:
1. Generate sample of the size N from multivariate normal distribution with parameters

$$\mu_1 = \begin{pmatrix} 0 \\ \dots \\ 0 \end{pmatrix}, \Sigma_1 = \begin{pmatrix} 1 & \dots & 0 \\ \vdots & \ddots & \vdots \\ 0 & \dots & 1 \end{pmatrix}$$

2. Estimate the covariance matrix $S_1 = \frac{1}{N-1} \cdot A_1$ by the generated data.
3. Generate sample of the size N from multivariate normal distribution with parameters

$$\mu_2 = \begin{pmatrix} 0 \\ \dots \\ 0 \end{pmatrix}, \Sigma_2 = \begin{pmatrix} 1 & \dots & 0 \\ \vdots & \ddots & \vdots \\ 0 & \dots & 1 \end{pmatrix}$$

4. Estimate the covariance matrix $S_2 = \frac{1}{N-1} \cdot A_2$ by the generated data.
5. Test the hypothesis $H_0 : \Sigma_1 = \Sigma_2$ using Box's M-test at the significance level $\alpha = 0.05$.
6. Repeat the procedure $S = 10000$ times, and estimate the probability of type I error $\hat{\alpha}$.

Results are presented in Table 1. Table 1 contains the following information: first column describes the number of observations, first row describes the dimension of covariance matrix, data of table are $\hat{\alpha}$ for conducted experiments.

The results, given in Table 1, allow to make a conclusion that for the sample size 250 (one year of observations of stock returns) and the number of stocks p larger than 30 the given significance level $\alpha = 0.05$ is not controlled by the Box's M-test. Therefore, in this paper, the experimental results are limited by consideration of 10 stocks.

Table 1 Estimation of type I error

N	p			
	10	30	50	100
150	0.0529			
200	0.0525	0.0893		
250	0.052	0.0777	0.1555	
500	0.0501	0.0715	0.0989	0.3108
1000	0.0498	0.0595	0.0761	0.2648
1500		0.0546	0.0654	0.1335
2000			0.0593	0.1028
3000				0.0853

5 Experimented Results

The described methodology is applied to test homogeneity hypothesis (1) by real data of stock returns of German and French stock markets over 12-years period from 2003 to 2014. Observations are divided into 12 parts by number of years, which determine 66 individual hypotheses (2) for German market and 66 individual hypotheses (2) for French market.

Companies with high share capitalization were selected: 10 German companies—from $DAX30$ index; 10 French companies—from $CAC40$ index. The lists of the German and French companies chosen for the study are given in Tables 2 and 3 correspondingly.

The results for German stock market are given in Table 4. The results for French stock market are given in Table 5.

Table 2 German companies with high share capitalization

	Ticker symbol	Company	Sector
1	ADS.DE	Adidas	Consumer Goods
2	ALV.DE	Allianz	Financial
3	BEI.DE	Beiersdorf	Consumer Goods
4	DPW.DE	Deutsche Post	Services
5	DTE.DE	Deutsche Telekom	Technology
6	LIN.DE	Linde	Basic Materials
7	MAN.DE	MAN SE	Consumer Goods
8	MRK.DE	Merck KGaA	Healthcare
9	SAP.DE	SAP	Technology
10	SIE.DE	Siemens	Industrial Goods

Table 3 French companies with high share capitalization

	Ticker symbol	Company	Sector
1	AF.PA	Air France-KLM	Services
2	STM.PA	STMicroelectronics	Technology
3	EAD.PA	EADS	Aerospace industry
4	UG.PA	PSA Peugeot Citron	Consumer Goods
5	SEV.PA	Suez Environnement	Utilities
6	ALU.PA	Alcatel-Lucent	Technology
7	ML.PA	Michelin	Consumer Goods
8	DX.PA	Dexia	Banks
9	RI.PA	Pernod Ricard	Consumer Goods
10	LG.PA	Lafarge	Construction Materials

Table 4 Results. Germany

	2003	2004	2005	2006	2007	2008	2009	2010	2011	2012	2013	2014
2003	0	1	1	1	1	1	1	1	1	1	1	1
2004	1	0	1	1	1	1	1	1	1	1	1	1
2005	1	1	0	1	1	1	1	1	1	1	1	1
2006	1	1	1	0	1	1	1	1	1	1	1	1
2007	1	1	1	1	0	1	1	1	1	1	1	1
2008	1	1	1	1	1	0	1	1	1	1	1	1
2009	1	1	1	1	1	1	0	1	1	1	1	1
2010	1	1	1	1	1	1	1	0	1	1	1	1
2011	1	1	1	1	1	1	1	1	0	1	1	1
2012	1	1	1	1	1	1	1	1	1	0	1	1
2013	1	1	1	1	1	1	1	1	1	1	0	1
2014	1	1	1	1	1	1	1	1	1	1	1	0

Table 5 Results. France

	2003	2004	2005	2006	2007	2008	2009	2010	2011	2012	2013	2014
2003	0	1	1	1	1	1	1	1	1	1	1	1
2004	1	0	1	1	1	1	1	1	1	1	1	1
2005	1	1	0	1	1	1	1	1	1	1	1	1
2006	1	1	1	0	1	1	1	1	1	1	1	1
2007	1	1	1	1	0	1	1	1	1	1	1	1
2008	1	1	1	1	1	0	1	1	1	1	1	1
2009	1	1	1	1	1	1	0	1	1	1	1	1
2010	1	1	1	1	1	1	1	0	1	1	1	1
2011	1	1	1	1	1	1	1	1	0	1	1	1
2012	1	1	1	1	1	1	1	1	1	0	1	1
2013	1	1	1	1	1	1	1	1	1	1	0	1
2014	1	1	1	1	1	1	1	1	1	1	1	0

Tables 4 and 5 contain the results of testing individual hypotheses

$$h_{ij} : \Sigma_i = \Sigma_j \ vs \ k_{ij} : \Sigma_i \neq \Sigma_j; i, j = 2003, \ldots, 2014$$

using Box's M-test. First column in the Tables 4 and 5 describes years $i = 2003$, ..., 2014, first row describes years $j = 2003$, ..., 2014, next columns describe results of hypothesis testing. In Tables 4 and 5 value 1 in the row i and column j means that the hypothesis $h_{i,j}$ is rejected at significance level $\alpha' = \frac{\alpha}{66} = \frac{0.05}{66} = 0.0007575$. The value 0 in the row i and column j means that the hypothesis $h_{i,j}$ is accepted at significance level $\alpha' = 0.0007575$.

Table 6 German companies belonging to chemical industry

	Ticker symbol	Company	Sector
1	BAS.DE	BASF	Chemical industry
2	BAYN.DE	Bayer	Chemical industry
3	HEN3.DE	Henkel AG and Co. KGaA	Chemical industry
4	LIN.DE	Linde	Chemical industry
5	LXS.DE	Lanxess AG	Chemical industry
6	SGL.DE	SGL Carbon	Chemical industry

It is possible to make a conclusion that for the German stocks described in Table 2 all hypotheses of stability of stock returns connections are rejected for any pair of years. This conclusion is also correct for French stocks described in Table 3.

It is shown that for the selected stocks there is no stability of connections of stock returns for different periods of observation. Note, that these companies belongs to different sectors. To continue our study we consider the stocks from the same sectors.

Germany. German companies belonging to chemical industry and existing in the market the whole period from 2003 to 2014 were selected. The list of companies is given in Table 6.

The results for German stocks described in Table 6 are given in Table 7. Table 7 contain the results of testing individual hypotheses

$$h_{ij} : \Sigma_i = \Sigma_j \ vs \ k_{ij} : \Sigma_i \neq \Sigma_j ; i, j = 2003, \ldots, 2014$$

using Box's M-test. First column in the Table 7 describes years $i = 2003, ..., 2014$, first row describes years $j = 2003, ..., 2014$, next columns describe results of hypothesis testing. The results of hypotheses testing with *p-values* are given in Table 7. If there is the value at the intersection of the *ith* row and *jth* column less than 0.0007575 then the hypothesis $h_{ij} : \Sigma_i = \Sigma_j$ is rejected at significance level $\alpha' = 0.0007575$; if this value is more than 0.0007575, the hypothesis $h_{ij} : \Sigma_i = \Sigma_j$ is accepted at significance level $\alpha' = 0.0007575$.

The experimented results in Table 7 allows to make a conclusion that the homogeneity hypotheses of covariance matrices: $\Sigma_{2006} = \Sigma_{2007}$ and $\Sigma_{2013} = \Sigma_{2014}$ are not rejected by multiple testing procedure. Similar studies for stocks of companies belonging to the sector of services were carried out. The list of the selected companies is presented in Table 8.

Table 7 Results. Germany

	2003	2004	2005	2006	2007	2008	2009	2010	2011	2012	2013	2014
2003	1	0	0	0	0	0	0	0	0	0	0	0
2004	0	1	0	0	0	0	0	0	0	0	0	0
2005	0	0	1	0	0	0	0	0	0	0	0	0
2006	0	0	0	1	0.00077	0	0	0	0	0	0	0
2007	0	0	0	0.00077	1	0	0	0	0	0	0	0
2008	0	0	0	0	0	1	0	0	0	0	0	0
2009	0	0	0	0	0	0	1	0	0	0	0	0
2010	0	0	0	0	0	0	0	1	0	0	0	0
2011	0	0	0	0	0	0	0	0	1	0	0	0
2012	0	0	0	0	0	0	0	0	0	1	0	0
2013	0	0	0	0	0	0	0	0	0	0	1	0.024
2014	0	0	0	0	0	0	0	0	0	0	0.024	1

Table 8 German companies belonging to the sector of services

	Ticker symbol	Company	Sector
1	EVD.DE	CTS Eventim AG and Co. KGaA	Services
2	FIE.DE	Fielmann	Services
3	AAD.DE	Amadeus Fire	Services
4	BVB.DE	Borussia Dortmund	Services
5	BYW6.DE	BayWa	Services
6	GFK.DE	GfK	Services
7	TTK.DE	TAKKT	Services

The results for German stocks described in Table 8 are given in Table 9. Table 9 contain the results of testing individual hypotheses

$$h_{ij} : \Sigma_i = \Sigma_j \ vs \ k_{ij} : \Sigma_i \neq \Sigma_j; i, j = 2003, \ldots, 2014$$

using Box's M-test. First column in the Table 9 describes years $i = 2003, ..., 2014$, first row describes years $j = 2003, ..., 2014$, next columns describe results of hypothesis testing. The results with test *p-values* for testing hypotheses $h_{i,j}$ are given in Table 9. If there is the value at the intersection of the *ith* row and *jth* column less than 0.0007575 then the hypothesis $h_{ij} : \Sigma_i = \Sigma_j$ is rejected at significance level $\alpha' = 0.0007575$; if this value is more than 0.0007575, the hypothesis $h_{i,j} : \Sigma_i = \Sigma_j$ is accepted at significance level $\alpha' = 0.0007575$.

The Table 9 shows that the hypotheses $\Sigma_{2006} = \Sigma_{2007}$ and $\Sigma_{2012} = \Sigma_{2013}$ are accepted by the multiple testing procedure.

There are sets of 10 companies for German stock market for which not all the homogeneity hypotheses are rejected. Such sets of companies are given in Tables 10 and 11. For the set of 10 companies described in Table 10 hypothesis $\Sigma_{2006} = \Sigma_{2007}$ is accepted by the multiple testing procedure. For the set of 10 companies described in Table 11 hypothesis $\Sigma_{2013} = \Sigma_{2014}$ is accepted by the multiple testing procedure.

France. There are sets of 10 companies for French stock market for which not all the homogeneity hypotheses are rejected. Such sets of companies are given in Tables 12 and 13. For the set of 10 companies described in the Table 12 hypotheses $\Sigma_{2004} = \Sigma_{2005}$ and $\Sigma_{2005} = \Sigma_{2006}$ are accepted by the multiple testing procedure. For the set of 10 companies described in Table 13 hypothesis $\Sigma_{2013} = \Sigma_{2014}$ is accepted by the multiple testing procedure.

Table 9 Results. Germany, sector of services

	2003	2004	2005	2006	2007	2008	2009	2010	2011	2012	2013	2014
2003	1	0	0	0	0	0	0	0	0	0	0	0
2004	0	1	0	3×10^{-7}	0.00048	0	0	0	0	0	0	0
2005	0	0	1	0	0	0	0	0	0	0	0	0
2006	0	3×10^{-7}	0	1	0.0197	0	0	0	0	0	0	0
2007	0	0.00048	0	0.0197	1	0	0	0	0	0	0	0
2008	0	0	0	0	0	1	0	0	0	0	0	0
2009	0	0	0	0	0	0	1	0	0	0	0	0
2010	0	0	0	0	0	0	0	1	0	0	0	0
2011	0	0	0	0	0	0	0	0	1	0	0	0
2012	0	0	0	0	0	0	0	0	0	1	0.03	0
2013	0	0	0	0	0	0	0	0	0	0.03	1	0
2014	0	0	0	0	0	0	0	0	0	0	0	1

Table 10 Germany companies

	Ticker symbol	Company	Sector
1	BAS.DE	BASF	Chemical industry
2	BAYN.DE	Bayer	Chemical industry
3	HEN3.DE	Henkel AG and Co. KGaA	Chemical industry
4	EVD.DE	CTS Eventim AG and Co. KGaA	Services
5	FIE.DE	Fielmann	Services
6	AAD.DE	Amadeus Fire	Services
7	BVB.DE	Borussia Dortmund	Services
8	BYW6.DE	BayWa	Services
9	GFK.DE	GfK	Services
10	TTK.DE	TAKKT	Services

Table 11 German companies

	Ticker symbol	Company	Sector
1	BDT.DE	Bertrandt	Auto Parts
2	DEZ.DE	Deutz AG	Diversified Machinery
3	CON.DE	Continental	Auto Parts
4	RAA.DE	Rational AG	Industrial Equipment Wholesale
5	GIL.DE	DMG Mori Seiki AG	Machine Tools and Accessories
6	G1A.DE	GEA Group	Diversified Machinery
7	AIR.DE	AIRBUS GROUP	Aircraft industry
8	SZG.DE	Salzgitter AG	Steel and Iron
9	RWE.DE	RWE AG	Diversified Utilities
10	JEN.DE	Jenoptik AG	Diversified Electronics

Table 12 French companies

	Ticker symbol	Company	Sector
1	ALU.PA	Alcatel-Lucent	Communication Equipment
2	BIG.PA	BigBen Interactive	Electronic Equipment
3	GEA.PA	Grenobloise d'Electronique et d'Automatismes SA	Business Equipment
4	GID.PA	Egide SA	Electronic Equipment
5	BELI.PA	Le Blier Societe Anonyme	Auto Parts
6	EO.PA	Faurecia S.A.	Auto Parts
7	ALHIO.PA	Hiolle Industries	Industrial equipment
8	CRI.PA	Chargeurs SA	Textile Industrial
9	BEN.PA	Bnteau S.A.	Recreational Goods, Other
10	FII.PA	Lisi SA	Aerospace/Defense—Major Diversified

Table 13 French companies

	Ticker symbol	Company	Sector
1	CNP.PA	CNP Assurances Socit anonyme	Life Insurance
2	CS.PA	AXA Group	Property and Casualty Insurance
3	ELE.PA	Euler Hermes Group SA	Property and Casualty Insurance
4	ACA.PA	Crdit Agricole	Money Center Banks
5	BNP.PA	BNP Paribas	Money Center Banks
6	GLE.PA	Societe Generale Group	Money Center Banks
7	BOI.PA	Boiron SA	Healthcare
8	COX.PA	Nicox SA	Healthcare
9	DGM.PA	Diagnostic Medical Systems S.A.	Healthcare
10	DIM.PA	Sartorius Stedim Biotech S.A.	Healthcare

6 Conclusion

In this paper, the problem of testing of stability of stock returns connections over time is considered from multiple hypotheses testing point of view. For individual hypotheses testing Box's M-test is applied. The obtained procedure is applied to testing stability of stock returns connections for German and French stock markets over period from 2003 to 2014. It is shown that stability hypothesis is rejected. However, there are sets of stock returns for which stability hypotheses of connections are not rejected for some period of observations.

Acknowledgements The work of Koldanov P.A. was conducted at the Laboratory of Algorithms and Technologies for Network Analysis of National Research University Higher School of Economics. The work is partially supported by RFHR grant 15-32-01052.

References

1. Anderson, T.W.: An Introduction to Multivariate Statistical Analysis, p. 721. T.W. Anderson, New York. Wiley (2003)
2. Boginski, V.: Statistical analysis of financial networks. In: Boginski, V., Butenko, S., Pardalos, P.M. (eds.) Computational Statistics and Data Analysis, vol. 48, issue no 2, pp. 431–443 (2005)
3. Box, G.E.P.: A general distribution theory for a class of likelihood criteria. Biometrika **36**, 317–346 (1949)
4. Lehmann, E.L.: Chapter 9: Multiple testing and simultaneous inference. In: Lehmann, E.L., Romano, J.P. (eds.) Testing Statistical Hypotheses, pp. 348–391. Springer, New York (2005)

Part III
Applications

Network Analysis of International Migration

Fuad Aleskerov, Natalia Meshcheryakova, Anna Rezyapova and Sergey Shvydun

Abstract Our study employs the network approach to the problem of international migration. During the last years, migration has attracted a lot of attention and has been examined from many points of view. However, very few studies considered it from the network perspective. The international migration can be represented as a network (or weighted directed graph) where the nodes correspond to countries and the edges correspond to migration flows. The main focus of our study is to reveal a set of critical or central elements in the network. To do it, we calculated different existing and new centrality measures. In our research the United Nations International Migration Flows Database (version 2015) was used. As a result, we obtained information on critical elements for the migration process in 2013.

1 Introduction

Migration is one of the fundamental processes in the society. Violent conflicts, income inequality, poverty, and climate change lead to large movements of people and thus shape the world population and influence the society considerably. Therefore, international migration is the issue of high importance, and new theories and policies are

F. Aleskerov (✉) · N. Meshcheryakova · A. Rezyapova · S. Shvydun
National Research University Higher School of Economics (HSE), Moscow, Russia
e-mail: alesk@hse.ru

N. Meshcheryakova
e-mail: natamesc@gmail.com

A. Rezyapova
e-mail: annrezyapova@gmail.com

S. Shvydun
e-mail: shvydun@hse.ru

F. Aleskerov · S. Shvydun
V.A. Trapeznikov Institute of Control Sciences of Russian Academy of Sciences (ICS RAS), Moscow, Russia

V.A. Kalyagin et al. (eds.), *Models, Algorithms, and Technologies for Network Analysis*, Springer Proceedings in Mathematics & Statistics 197,
DOI 10.1007/978-3-319-56829-4_13

needed to be developed in order to contribute to the development of both home and host countries.

Migration was studied in various fields of science. A considerable amount of works was proposed in order to explain the causes of migration flows and the consequences of them. The first studies were focused on movements of people from rural to urban areas [1] and provided the fundamental understanding of factors influencing the migration [2]. Lately, several models from other areas were adapted to the study of migration process. A gravity model of migration plays a significant role in studying migration flows. The main hypothesis was that the level of migration between two territories is positively related to the population of them and inversely related to the distance between them [3, 4]. Several works explore the phenomenon of migration from the prospect of motives to migrate. There was an attempt to explain the migration process by push–pull factors [5], prospect of the economic theory and human capital approach [6, 7]. All these theories apply different levels of analysis of human migration: the macro-level (migration between countries and regions) and micro-level (individual). However, they lack the fact that migration is a complex process and level of migration between any two countries depends not only on factors related to these two countries, but also on migration flows between other countries.

The migration process can be also studied from the network perspective. The international migration can be represented as a network (or weighted directed graph) where the nodes correspond to countries and the edges correspond to migration flows. This approach allows to consider the flows between any two countries integrated into the whole system of countries and shows how the changes in one flow may affect the flows between the other seemingly unrelated countries. The application of the network approach to the international migration was presented in [11–13]. Unfortunately, these studies did not fully take into consideration nodes attributes, individual and group influences of nodes.

Our work is aimed to detect the countries with highest level of importance in the international migration network. For this purpose we evaluate the classical and new centrality indices. Classical centrality indices are essential for the representation of major migration flows while indices developed in [11–13] take into account the node attributes population of the destination country as well as indirect connections and group influence between the countries in the network.

The paper is organized as follows. First, we provide the main information on centrality indices. Second, we describe the data, evaluate the indices and then give an interpretation of the main results.

2 Centrality Measures

In our work the following known centrality measures were evaluated: degree and weighted degree centrality, closeness, eigenvector, and PageRank.

The degree centrality is the number of nodes each node is connected with [14]. For directed graph the degree centrality has three forms: the degree, indegree and outdegree centrality. The indegree centrality represents the number of incoming ties each node has, and outdegree is the number of outgoing ties for each node.

In terms of migration, edge in unweighted graph characterizes the presence of migration flow between any two countries. The indegree centrality for country A is the number of countries, which are connected with country A through migration inflows to country A. In other words, it is the number of countries, which migrants came to country A from. For outdegree the interpretation is as follows: the number of countries which are connected with country A through migrant outflows from A or the number of countries which are the destinations of migrants from A. The degree centrality of country A can show how many different countries are connected with it through migration flows.

The following centrality indices were estimated for the weighted network: weighted in-degree, weighted outdegree, weighted degree difference (=weighted in-degree weighted outdegree), and weighted degree [14]. *Weighted indegree* (WIn) centrality represents the number of incoming ties for each node with weights on them or the immigrant flow to the country. *Weighted outdegree* (WOut) is the number of outgoing links for each node and accordingly relates to the number of emigrants. *Weighted degree difference* (WDegD) is the difference between migrant inflow and outflow which is the net migration flow. *Weighted degree* (WDeg) is the sum of weighted indegree and weighted outdegree centralities for each country or the total number of emigrants and immigrants (gross migration). These centrality indices can give us the basic information about the international migration process: the level of migrant inflows and outflows, net and gross migration flows.

The *closeness* (Clos) [15] centrality shows how close node is located to the other nodes in the network. In addition, this measure has the following characteristics. First, it accounts only for short paths between nodes. Second, these centralities have very close values and are sensitive to the changes in network structure: minor changes in the structure of network can lead to significant differences in ranking by this measure. In our work the closeness centrality is estimated for the undirected graph with maximization of the weights on paths and is related to the level of closeness of particular country to intense migration flows. Note that it does not imply that the country itself should have huge migration inflows or outflows. This measure can provide the information about potential migration flow to particular country by estimating the distance between the country and countries with huge migration flows in the network. Countries with low closeness centrality value are not necessarily involved in the process of international migration since they usually have low migration flows.

Eigenvector (EV) [16] is the generalized degree centrality, which accounts for degrees of node neighbors. Eigenvector centrality and its counterpart *PageRank* (PR) [17] centrality measure are based on the idea that a particular node has a high importance if its adjacent nodes linked have a high importance. In international migration network these indices highlight the countries' centers of international immigration, and the countries, which are directly linked with them through migration flows.

Unfortunately, classical indices do not elucidate hidden elements influential in the network. This can be explained by the fact that these indices do not fully take into account individual properties of nodes, the intensity level of direct connections and interactions between nodes of the networks.

In [12] a novel method for estimating the intensities of nodes interactions was proposed. The method is based on the power index analysis that proposed in [11] and is similar to the Banzhaf index used to evaluate the power of each agent in voting procedures [18]. The index (originally called a key borrower index) is a Short-Range Interaction Centrality (SRIC) that was adjusted for the network theory and employed to find the most pivotal borrower in a loan market in order to take into account some specific characteristics of financial interactions. Unfortunately, SRIC index do not take into account long-range interactions in the network, which leads to the fact that nodes that are not adjacent do not influence each other. To overcome the shortage of SRIC index several Long-Range Interaction Centrality (LRIC) indices were proposed in [13].

We evaluate the direct influence of one country to another one through imposing the quota, which represents the population of the destination country. We suggest that 0.1% of population of destination country is the critical level of migrant inflow. If the migration flow from country A to country B does not reach 0.1% of population of country B, then country A does not directly influence country B through migration flows.

To sum up, the classic centrality indices and indices of Short- and Long-Range Interactions are applied to characterize the countries in migration network. The distinctive feature of the latter is the consideration of the population of destination country and indirect migration routes between countries.

3 Key Actors in International Migration Network

3.1 Data Description

Data on international migration is usually presented in two fundamental statistical categories: stock of migrants and migration flows. Migration flow is defined as a number of persons arriving to country or leaving it in a given time period. Migrant stock corresponds to the total number of people living in a country other than the country of origin in a certain moment. The key difference between these two categories is that the stock of migrants is an accumulative pattern, and the flow data represents the fact of immigration or emigration to or from a given country.

We use the data on dyadic migrant flow for analysis of the international migration provided by the United Nations [19]. The list of responded actors in the database contains 45 countries. Migration flows for countries not included in the list were accumulated by the statistics of the countries presented in the database. The data was collected through different sources: population registers, border statistics, the number

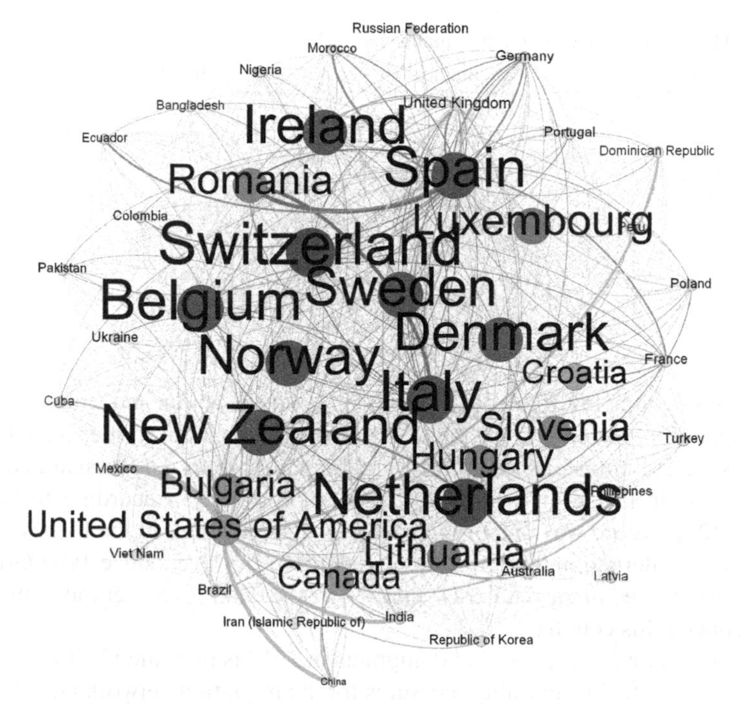

Fig. 1 The international migration network for 2013

of residents permits issued, statistical forms that persons fill when they change place of residence and household survey. There are three ways to define country of migrants origin or destination by

(1) residence; (2) citizenship; (3) place of birth.

Additionally, as countries apply different criteria to determine international migrant and the country of origin, collect data through different sources and have various purposes of migration policy, there were some cases of inconsistency in observations. Thus, there were several techniques performed of data aggregation to resolve the problems of inconsistency in observations.

The international migration network for 2013 is shown on Fig. 1.

3.2 *Influence in International Migration Network*

As it was mentioned before we applied different centrality indices to provide a ranking and detect the most influential countries in migration process. The centrality indices are evaluated for 2013 which is the last year in the database. The major international migration flows occurred between the following groups of countries. First, the migration flow from Mexico, the Philippines, and Vietnam to the USA were

Table 1 Migration flows over 50,000 migrants in 2013

Origin	Destination	Migration flow
Mexico	USA	135,028
China	USA	71,798
Spain	Romania	70,055
India	USA	68,458
Romania	Italy	59,347
Philippines	USA	54,446

still of considerable level. Second, new Asian countries, India and China, appeared among labor force suppliers for the USA. Flows between the former Soviet Union countries were diminishing after 2007, and migration from the Russian Federation and Kazakhstan to Germany was decreasing accordingly. According to Eurostat statistics [20] Greece was one of the countries that experienced the highest growth in number of international migrants in recent time. However, since 1998 Greece is no longer presented in the databases, that is why the rankings by centrality measures do not contain this country.

Information on flows over 50000 migrants in 2013 is presented in Table 1.

Let us calculate the centrality measures for the migration network (see Table 2).

From the results for weighted indegree centrality we can conclude that the highest number of immigrants were received by the USA, Italy, and the UK. According to the ranking by weighted outdegree, Spain, India, and China had the highest migrant outflow. Weighted degree ranking highlights the USA, Spain, Italy and the UK, which had the greatest gross migration rate. The weighted degree difference or the highest net migration flow was in the USA, Canada, the UK, and Italy.

Different results can be obtained from the estimation of the level of closeness: the USA is still the first, however, Mexico, Netherlands, Spain, and Switzerland are presented. These countries had intense migration inflows (the USA) or outflows (Spain) itself, or had migration flows to or from the countries with intense migration [21]. Mexico–US migration route was established historically, and now Mexicans are accounted for 28% of foreign-born population in the USA [22]. Netherlands and Switzerland were connected through migration flows to Italy, which was the second immigration country after the USA.

Eigenvector and PageRank highlight the rich-club group of countries: the USA, Italy, the UK, and Spain. These countries are involved in the process of migration more than others and in addition had flows between each other. In this case eigenvector and PageRank centralities can show how mobile is the population of countries.

Ranking by classic centrality indices provided us with the information about countries with the highest in- and outflows of migrants, net migration flow, level of closeness to huge migration flows and countries most involved in migration process. Short-Range and Long-Range Interaction Centralities can help us to explore the international migration network from the different perspective.

Table 2 Rankings by centrality indices for 2013

Country	WIn	WOut	WDeg	WDegD	Clos	PR	EV	SRIC	LRIC (SUM)	LRIC (MAX)	LRIC (MAXMIN)
USA	1	19	1	1	1	1	2	22	6	10	10
Italy	2	5	3	4	6	6	4	11	10	11	16
UK	3	10	4	3	30	3	1	9	9	4	7
Canada	4	44	5	2	10	7	12	74	37	43	30
Spain	5	1	2	215	3	2	3	1	1	1	1
Switzerland	6	12	7	6	5	5	6	35	44	54	80
Netherlands	7	8	8	10	4	8	11	17	14	23	27
Sweden	8	21	15	5	9	11	19	15	30	38	35
Belgium	9	14	10	9	7	12	9	23	19	28	45
Romania	10	6	6	198	14	17	5	2	2	2	2
Germany	11	11	9	23	37	10	7	12	4	8	9
New Zealand	12	16	13	14	8	4	14	5	23	15	15
France	13	9	12	192	36	15	8	7	3	5	5
Norway	14	52	23	7	11	16	23	32	45	49	24
Australia	15	31	22	8	33	9	20	18	21	25	21

Spain, Romania, India, and Poland had the highest ranks according to the index of Short-Range Interaction Centrality. These results are highly related to the weighted outdegree. Additionally, SRIC accounts for the first-order indirect interactions and the population of destination country. That is why there was a little change in the order of countries with intense emigration flows.

Three of LRIC indices show almost similar results: Spain, Romania, France, Germany, Poland, and India are at the top of rankings. Spain has the highest emigration rate. Romania, India, and France have the migration flows to countries with huge population and intense migration flows. There was a huge flow from India to the USA, the USA has large population and is a popular country of migrants' destination [22]. France is presented in ranking by LRIC indices, because it has migration flows to Spain (10,548) and to the UK (24,313). Romania also had migration flows to the UK. Poland did not appear among countries with highest emigration rate (weighted outdegree), however, it had migration flow of almost 10000 migrants to Norway with population of around 5 million people. The share of this migrant inflow (0.2%) exceeded 0.1% of the population of Norway. This result is important to be considered as when migration flow is more than level expected by the destination country, it can lead to negative consequences for both migrants and the population of destination country.

The results introduced by classical centralities and SRIC, LRIC indices both outline the emigration countries. However, SRIC and LRIC indices introduce additionally the emigration countries with considerable for the population of destination country share of migrants (Poland).

4 Conclusion

International migration studied from different points of view. Our study was focused on the network analysis of migration process to determine the most influential actors. Estimation of classical centrality indices is the one of the possible ways to analyze countries influence in the network through migration flows. Our work goes a step further and allows to consider indirect connections of countries in the international migration network and a node attribute the population of destination country. This idea is implemented through Short-Range and Long-Range Interaction Centralities.

The analysis was applied to data on migration flows in 2013. Our methodology outlined not only the countries with large number of immigrants or emigrants, but also the countries with migrant outflows considerable for the population of destination country and emigration to the popular destination countries. These results are important in order to provide countries highly involved in the process of international migration with relevant migration policy.

Acknowledgements The paper was prepared within the framework of the Basic Research Program at the National Research University Higher School of Economics (HSE) and supported within the framework of a subsidy by the Russian Academic Excellence Project '5–100'. The work was conducted by the International Laboratory of Decision Choice and Analysis (DeCAn Lab) of the National Research University Higher School of Economics.

References

1. Smith, A., Garnier, M.: An Inquiry into the Nature and Causes of the Wealth of Nations. Thomas Nelson, Edinburgh (1838)
2. Ravenstein, E.G.: The laws of migration. J. Roy. Stat. Soc. **52**(2), 241–305 (1889)
3. Tinbergen, J.: Shaping the world economy: suggestions for an international economic policy. HD82 T54. The Twentieth Century Fund, New York (1962)
4. Zipf, G.K.: The P1 P2/d hypothesis: on the intercity movement of persons. Am. Sociol. Rev. **11**(6), 677–686 (1946)
5. Lee, E.: A theory of migration. Demography **3**(1), 47–57 (1966)
6. Bodvarsson, B., Simpson, N.B., Sparber, C.: Migration Theory. Handbook of Economics of International Migration. The Immigrants, vol. 1, 1st edn. Elsevier Inc. (2015)
7. Sjaastad, L.: The costs and returns of human migration. J. Polit. Econ. **70**, 80–93 (1962)
8. Davis, K., D'Odorico, P., Laio, F., Ridolfi, L.: Global spatio-temporal patterns in human migration: a complex network perspective. PLoS One, **8**(1), e53723 (2013)
9. Fagiolo, G., Mastrorillo, M.: The International-Migration Network (2012). eprint arXiv:1212.3852
10. Tranos, E., Gheasi, M., Nijkamp, P.: International migration: a global complex network. Environ. Plan. **42**(1), 4–22 (2015)
11. Aleskerov, F.T.: Power indices taking into account agents preferences In: Simeone, B., Pukelsheim, F. (eds) Mathematics and Democracy, pp. 1–18. Springer, Heidelberg (2006)
12. Aleskerov, F., Andrievskaya, I., Permjakova, E.: Key Borrowers Detected by the Intensities of Their Short-Range Interactions, Higher School of Economics Research Paper (2014)
13. Aleskerov F.T., Meshcheryakova N.G., Shvydun S.V.: Centrality measures in networks based on nodes attributes, long-range interactions and group influence. In: Series WP7 Mathematical

methods for decision making in economics, business and politics, pp. 1–44. HSE Publishing House, Moscow, No. WP7/2016/04 (2016)
14. Freeman, L.C.: Centrality in social networks: conceptual clarification. Soc. Netw. **1**, 215–239 (1979)
15. Bavelas, A.: Communication patterns in task-oriented groups. J. Acoust. Soc. Am. **22**(6), 725–730 (1950)
16. Bonacich, P.: Technique for analyzing overlapping memberships. Sociol. Methodol. **4**, 176–185 (1972)
17. Brin, S., Page, L.: The anatomy of a large-scale hypertextual Web search engine. Comput. Netw. **30**, 107–117 (1998)
18. Banzhaf, J.F.: Weighted voting doesn't work: a mathematical analysis. Rutgers Law Rev. **19**, 317–343 (1965)
19. United Nations, Department of Economic and Social Affairs, Population Division. International Migration Flows to and from Selected Countries: The 2015 Revision (POP/DB/MIG/Flow/Rev.2015) (2015). Accessed 30 May 2016
20. http://ec.europa.eu/eurostat/statistics-explained/index.php. Accessed 28 May 2016
21. United Nations Development Program. Human Development Report – 2009 Overcoming barriers: Human mobility and development. Human Development, vol. 331 (2009)
22. http://www.migrationpolicy.org/article/mexican-immigrants-united-states. Accessed 20 Apr 2016

Overlapping Community Detection in Social Networks with Node Attributes by Neighborhood Influence

Vladislav Chesnokov

Abstract Community detection is one of the key instruments in social network analysis. In social networks nodes (people) have many attributes such as gender, age, hometown, interests, workplace, etc., which can form possibly overlapping communities. But quite often full information about a person's attributes cannot be obtained due to data loss, privacy issues, or person's own accord. A fast method for overlapping community detection in social networks with node attributes is presented. The proposed algorithm is based on attribute transfer from neighbor vertices, and does not require any knowledge of attributes meaning. It was evaluated on Facebook and Twitter datasets with ground-truth communities and four classic graphs: Zachary's karate club, books about US politics, American college football, and US political blogs. Also, author's ego-network with manually labeled communities from VKontakte was used for the evaluation. Experiments show that the proposed method outperforms such algorithms as Infomap, modularity maximization, CESNA, BigCLAM, and AGM-fit by F_1-score and Jaccard similarity coefficient by more than 10%. The algorithm is tolerant to node attributes partial absence: more than 50% of attributes values can be deleted without great loss in accuracy. It has a near-linear runtime in the network size and can be easily parallelized.

1 Introduction

Community detection in graphs is an important task in many fields of science: biology, sociology, computer sciences, and others. In social graphs, vertices are people, and edges are connections between them: family ties, friendship, working relations, etc. The explosive growth of online social networking services, such as Facebook, Twitter and VKontakte, and open access to a huge amount of users' personal data gave an opportunity for many social researches: social groups detection for targeted advertisement, person's credit status obtainment by social networks profiles analysis,

V. Chesnokov (✉)
Bauman Moscow State Technical University, ul. Baumanskaya 2-ya, 5, Moscow, Russia
e-mail: v.o.chesnokov@yandex.ru

V.A. Kalyagin et al. (eds.), *Models, Algorithms, and Technologies for Network Analysis*, Springer Proceedings in Mathematics & Statistics 197,
DOI 10.1007/978-3-319-56829-4_14

detection of public opinion leaders for misinformation spread or usage in information warfare [1], and many others.

Classic graph clustering methods, such as k-means, modularity maximization, or the ones based on minimum spanning tree graph, are basically graph partition on nonintersecting subgraphs, clusters. They use only information about graph structure, i.e. its edges. But real social networks are not only nodes representing people with edges showing connections between them. Every person has a set of attributes like gender, age, education, hometown, favorite music, and so on. Algorithms which use node attributes information as well as graph structure just started to appear recently. They include Circles [2], CODICIL [3], and some others [4].

Besides, each node does not necessarily belong to a single community. Each person naturally has many activities: work, friends, hobbies, etc., so he or she can be a member of several communities, and therefore communities can overlap or be nested in social networks. Only few algorithms define and detect communities in such a way [5].

Only recently have approaches for detecting overlapping communities based on both sources of information been developed. An algorithm which uses a cell-based subspace clustering approach and identifies cells with dense connectivity in the subspaces is proposed in [6]. However, it requires knowledge about attribute nature and their filtering, because it uses reduced multi-valued attribute space. There are several methods based on topic models, e.g., [7], but most of them are quite slow and have poor scalability. EDCAR [8] algorithm relies on clique-based heuristics, but also requires high amount of computations. CESNA algorithm presented in [9] is based on probabilistic generative model. Yang et al. showed that this approach outperforms other state-of-art algorithms such as Block-LDA [7], CODICIL [3], and EDCAR [8] on real datasets from online social networks. Recently, an algorithm based on joint nonnegative matrix factorization approach was developed [10]. It gives almost the same F-scores of community detection as CESNA. FCAN algorithm [11], based on maximization of content relevance measure of vertices in cluster, outperforms CESNA on synthetic datasets. But CESNA still gives higher or almost the same NMI and accuracy scores on datasets for real networks of users from Facebook and Twitter [11]. However, it is still a challenge for an algorithm to detect human-labeled communities.

People tend to have ties with other people they share something common with [12, 13]. It was shown that vertex attributes can be inferred from its neighbors information [14–16]. This is a core idea of the proposed algorithm for overlapping community detection. The developed method uses both sources of information: node attributes as well as edge structure. Also it automatically labels obtained communities, i.e., each detected community comes with a set of attributes which probably formed it. The algorithm was originally developed for detecting communities in ego-networks. However, it can also be applied to any graph and gives quite good results on them.

2 Problem Definition

Consider unweighted undirected graph $G'(V', E')$ with $D(G') = 2$ which has $u \in V'$ such that

$$\forall v \in V', v \neq u\ \exists\{u, v\} \in E', \tag{1}$$

where $D(G')$ is the diameter of graph G'. Then it is called an ego-network for node u. Denote

$$V = V' \setminus \{u\}, \tag{2}$$

$$E = E' \setminus \{\{u, v\} | v \in V\}. \tag{3}$$

Each vertex from V' has a set of attributes (features) from set F, i.e., there exists a map $f : V' \rightarrow 2^F$. Obviously, one cannot always obtain full information about node features from real networks due to data transfer errors, censorship, privacy issues, etc. Denote by $f' : V' \rightarrow 2^F$ "real" or observed part of f such that

$$\forall v : f'(v) \subseteq f(v). \tag{4}$$

The task of community detection is to find a cover of vertex set V from graph $G(V, E)$ with observed f', such that "similar" vertices belongs to the same set(s).

3 The Proposed Algorithm

Previously [17], an algorithm for social graph clustering based on node attribute transfer was described. An overlapping community detection method based on similar principles is described in the current paper. The proposed algorithm is based on three assumptions. First, social networks have triadic structure [18], i.e., if there are edges between nodes A and B and A and C, then there is a link between B and C with high probability. Second, social networks obey affiliation model [19, 20], i.e., links between vertices are formed by communities and not vise versa [9]. That means that if two nodes belong to the same community or communities, they have one or more common attributes. Finally, vertex attributes can be inferred from node attributes of its neighborhood [14–16].

The algorithm has five main steps. At first, cores of communities are obtained. If some vertex belongs to community where all vertices have attribute a, then, obviously, the majority of its neighbors has this attribute. They can even form a clique. These attributes are called key attributes from now on. So, the algorithm finds such attributes for each vertex. Since the information about them may be partly missing, the procedure is iterative and both "normal" and key attributes are considered when the majority is determined. The communities grow with each iteration until they fill "dense" part of the graph. Note that some attributes are automatically filtered out, because not all attributes can form a dense subgraph, e.g., birth date or first name.

More formally, for each vertex v from V initialize an empty set of key attributes K_v. All nodes are visited at each iteration of the first step. For each vertex v a set of its neighbors $\mathcal{N}_v$ is determined and for each attribute $a \in F$ sets

$$\mathcal{N}_{v,a} = \{w | w \in \mathcal{N}_v \land a \in f'(w)\} \tag{5}$$

and

$$Q_{v,a} = \{w | w \in \mathcal{N}_v \land a \in K_w\} \tag{6}$$

are computed. If sum $|\mathcal{N}_{v,a}| + |Q_{v,a}|$ is greater than the threshold

$$t_{v,a} = \max(2, \alpha_a |\mathcal{N}_v|), \tag{7}$$

where α_a is a fraction defining qualified majority for attribute a, then attribute a is added to the set of key attributes K_v. Note that vertices that had attribute a both beforehand and in set of their key attributes are counted twice. So, node attributes are transfered from vertex to vertex through the sets of key attributes. The step ends when there were no changes at the last iteration.

Some nodes can belong to an intersection of several communities, i.e. they belong to two or more communities at the same time. They do not have attributes of these communities in their key attributes sets because their neighbors do not form a majority.

The second step is a procedure of determining key attributes for these "boundary" vertices. Like the previous step, this one is iterative. In this case, the rule for adding an attribute to the key attributes set should be weakened, so the threshold is changed. Now it depends on the number of communities in vertex neighborhood

$$t'_{v,a} = \max(2, \alpha_a \frac{|\mathcal{N}_v|}{|\bigcup_{w \in \mathcal{N}_v} K_w \setminus K_v|}). \tag{8}$$

Also, only value $|Q_{v,a}|$ instead of sum $|\mathcal{N}_{v,a}| + |Q_{v,a}|$ is now considered, because we only interested in key attributes, which represent communities. This step also ends when there were no changes at the last iteration.

Now each vertex has a key attributes set, and each attribute in it (probably) corresponds to some community. But several communities can have some common attribute(s). For example, the ego of ego-network can take extra class in his high school and have some friends who also visit it. The ego also has classmates. The classmates and extra class friends have a common attribute (high school) but belong to different communities. So, the communities should be split into connected components.

The third step is communities obtaining and their partition. Vertices are joined into communities by key attributes, i.e., for each attribute a from $\bigcup_{v \in V} K_v$ a community

$$C_a = \{v | a \in K_v\} \tag{9}$$

is formed. Then each community is partitioned into connected components in the subgraph of corresponding attribute. Trivial parts which consist of one or two vertices are filtered out, because each pair of people in social networks can have an unique tie connecting them.

The fourth step is merging. If any two parts from the previous step consist of the same nodes, they are merged into one community. The union of parts' attributes is associated with this community. This step is helpful for further analysis of detected communities: for each community there is a set of attributes which (probably) formed it.

Some vertices may have an empty set of key attributes after communities were obtained. This could mean that they are connected by some unknown or unspecified attribute. The last step is an attempt to detect communities formed in such a way. Slightly modified steps one and three are applied to the subgraph which consists of vertices with empty key attributes set. All vertices in this subgraph have single attribute—"unknown". Communities obtained at this step are appended to the ones obtained in the previous step.

3.1 Complexity and Scalability

The algorithm has near-linear runtime. At the first step, all vertex neighbors are considered at most once at each iteration. All vertices are visited, so each edge is considered at most twice—at most once for each incident node. So, the complexity of each iteration is linear in the number of edges and vertices: $O(|V| + |E|)$. Each attribute can "travel" the longest path in graph during transfer through the key attributes sets, so the number of iterations can be estimated as $O(D(G))$ in the worst-case. Due to small world phenomenon, the number of iterations is small: during the experiments it was less than 10 for most of graphs.

The procedure of the second step is practically the same, so complexity of each of its iterations is also linear in graph size. At the third step communities are formed by key attributes sets. This can be done in $O(|V|)$ using a hash-table for each attribute. We consider each node only once and add node to hash-tables of all key attributes of this node. Splitting community into connected components can be done with simple breadth-first search in $O(|V| + |E|)$.

Filtering connected components by size and merging duplicates is a trivial task and can be done in linear time in components count. It is obvious that this count is less than V. The complexity of the last step is the same or less than the sum of complexities of other steps, so it is also near-linear. Summing up, the algorithm has near-linear runtime.

The proposed algorithm can be parallelized in two dimensions: attributes and nodes. At the first step, all attributes are updated independently, so there may be as many threads as attributes. Because the algorithm is iterative, vertices can also be considered independently, but this requires shared memory. MapReduce approach can be used to implement this algorithm.

3.2 Heuristics

Fractions α_a which define qualified majority for attributes $a \in F$ can be calculated by heuristic if some information about node attributes meaning is known. This information can be obtained through machine learning or manual definition. The simplest heuristic requires zero knowledge about attributes meaning: each α_a equals some fixed value in range between 0 and 1.

4 Experimental Evaluation

4.1 Metrics

To evaluate the quality of the proposed algorithm, a set of graphs with known human-labeled ground-truth communities was used (see Sect. 4.2). To estimate the similarity between obtained communities and the ground-truth communities, an approach described in [19] and later adopted in [9] was applied. The most similar ground-truth community was found for each obtained community C_i and similarity measure was averaged by all obtained communities. Also, the most similar obtained community was found for each ground-truth community C_j^* and similarity measure was averaged by all ground-truth communities. Then, the average of these two values was taken

$$\frac{1}{2|C^*|}\sum_{C_j^* \in C^*} \max_{C_i \in C} \delta(C_j^*, C_i) + \frac{1}{2|C|}\sum_{C_i \in C} \max_{C_j^* \in C^*} \delta(C_j^*, C_i). \tag{10}$$

Two metrics were used as similarity measure $\delta(C_i^*, C_j)$: F_1-score

$$F_1 = \frac{2 \cdot \text{precision} \cdot \text{recall}}{\text{precision} + \text{recall}} = \frac{2 \cdot |C_j^* \cap C_i|}{|C_j^*| + |C_i|}, \tag{11}$$

where

$$\text{precision} = \frac{|C_j^* \cap C_i|}{|C_i|}, \tag{12}$$

$$\text{recall} = \frac{|C_j^* \cap C_i|}{|C_j^*|}, \quad (13)$$

and Jaccard similarity

$$J = \frac{|C_j^* \cap C_i|}{|C_j^* \cup C_i|}. \quad (14)$$

4.2 Datasets

Two datasets available from the Stanford Large Network Dataset Collection [21] were used to evaluate the proposed algorithm: ego-networks of Facebook and ego-networks of Twitter with node attributes and ground-truth communities. Their stats are presented in the Table 1. Also, four classic graphs were used: Zachary's karate club [22], books about US politics [23], American college football [24] and US political blogs [25] (see Table 2). These graphs had no attributes, so the information about vertex ground-truth communities was used as vertex attributes to demonstrate algorithms performance on well-studied graphs when full information about nodes' attributes is available.

Besides, the author's ego-network from VKontakte online social networking service was used for tests. It has 144 vertices, 1121 edges, 1526 different attributes, and 11 manually labeled overlapping communities. The attributes represent places of middle and high education, workplaces information, and membership in groups (virtual communities). The graph and its communities are depicted in the Fig. 1. Most

Table 1 SNAP datasets description. N—total number of nodes, E—total number of edges, C—total number of communities, K—total number of node attributes, S—average community size, A — average number of communities per node

Dataset	N	E	C	K	S	A
Facebook	4 089	170 174	193	175	28.76	1.36
Twitter	15 120	2 248 406	3 140	33 569	15.54	0.39

Table 2 Classic graphs description. N—number of nodes, E—number of edges, C—number of communities

Graph	N	E	C
Zachary's karate club	34	78	2
Political books	105	441	3
American college football	115	613	12
Political blogs	1 490	19 090	2

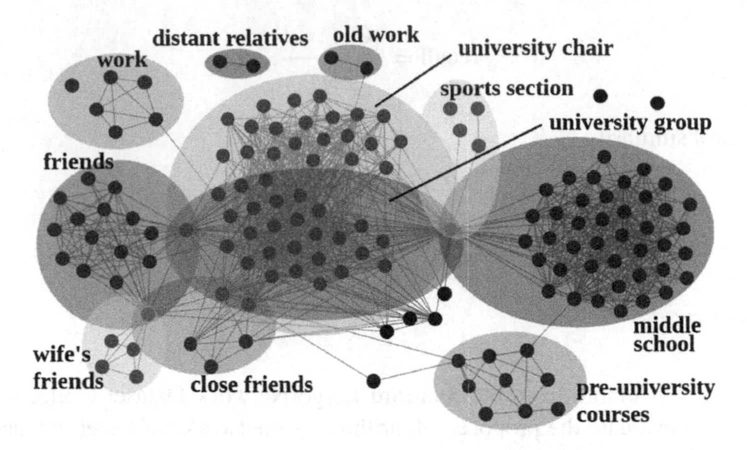

Fig. 1 The author's ego-network with human-labeled communities

vertices belong to one community, seven do not belong to any, five belong to three or more. One of the communities completely includes another.

4.3 Comparison to Other Algorithms

The proposed method was compared with algorithms Infomap [26], modularity maximization (Clauset-Newman-Moore variant [27]), AGM-fit [20], BigCLAM [19] and CESNA [9]. The first and the second algorithms are classic ones. They do not use information about node attributes and only detect nonoverlapping communities. The latter three were chosen because all of them rely on affiliation model. Implementations from SNAP framework [28] with default settings were used. AGM-fit and BigCLAM use only information about graph edges. In papers [19, 20] it was shown that these methods outperform Link Clustering [29], Clique Percolation [30], and Mixed-Membership Stochastic Block Mode [31] algorithms. CESNA uses both information from edge structure and node attributes and showed higher F_1 and Jaccard similarity scores than Demon [32], MAC [33], Block-LDA [7], CODICIL [3], EDCAR [8] and Circles [2] algorithms on Facebook and Twitter ego-networks datasets. Currently one can consider CESNA as the best algorithm which detects overlapping communities is social networks with node attributes.

The simplest heuristic, $\alpha_a = \beta$, was used in the proposed algorithm. The best score was taken for β in $[0.1, 0.9]$ with step 0.05 for each graph. For Facebook and Twitter datasets, scores were averaged for all algorithms. The last step of the proposed method was omitted for Twitter dataset.

The evaluation results are presented in Tables 3 and 4. The average F_1-score of the proposed method is higher than modularity maximization score by more than 12% on Facebook dataset. The proposed algorithm also outperforms BigCLAM by

Table 3 F_1-scores of AGM-fit, BigCLAM, CESNA, and the proposed algorithm. The average values for Facebook and Twitter datasets are shown

	Infomap	Modularity maximization	AGM-fit	BigCLAM	CESNA	Proposed algorithm
Facebook	0.414	0.484	0.466	0.447	0.447	0.543
Twitter	0.304	0.316	0.347	0.371	0.362	0.439
Zachary's karate club	0.510	0.789	0.663	0.399	0.431	0.970
Political books	0.294	0.629	0.671	0.340	0.594	0.809
American college football	0.790	0.608	0.479	0.871	0.820	0.917
Political blogs	0.164	0.550	0.370	0.401	0.327	0.811
Author's graph	0.631	0.792	0.706	0.624	0.656	0.843

Table 4 Jaccard index of AGM-fit, BigCLAM, CESNA, and the proposed algorithm. The average values for Facebook and Twitter datasets are shown

	Infomap	Modularity maximization	AGM-fit	BigCLAM	CESNA	Proposed algorithm
Facebook	0.312	0.383	0.366	0.325	0.339	0.442
Twitter	0.222	0.228	0.251	0.270	0.264	0.342
Zachary's karate club	0.361	0.680	0.497	0.252	0.279	0.942
Political books	0.179	0.533	0.536	0.207	0.436	0.742
American college football	0.679	0.459	0.354	0.801	0.729	0.887
Political blogs	0.102	0.476	0.227	0.251	0.196	0.685
Author's graph	0.518	0.721	0.674	0.496	0.566	0.795

average F_1-score by more than 18% on Twitter dataset. The results for Jaccard index are similar.

It also leaves all algorithms far behind by Zachary's karate club, books about US politics, and US political blogs F_1-scores, and shows results close to one. The proposed algorithm also shows high results on American college football and outperforms BigCLAM by 5% and CESNA by 12%. As for the author graph, it outperforms modularity maximization by 6%. The similar situation holds for the Jaccard index values. The developed algorithm outperforms others at least by 10%.

The developed algorithm gives quite high F_1-scores for classic graphs, but not as high as one might expect. One reason to that may be the fact that in real networks communities are not so dense as expected. Also, the task of community detection can be challenging even for "ego" of ego-network, not to mention other people, so for the same graph there can be different results depending on analyst.

4.4 An Example of Automatic Labeling

The detected communities and labels given for them by the algorithm for the author's graph are shown on Fig. 2. Four communities corresponding to author's middle school classmates, friends, pre-university courses acquaintances and author's wife's friends are detected perfectly, vertex to vertex. For the school classmates community, the algorithm gives a label consisting of three attributes: school number, school old number, and official group (virtual community) of school. The school number was changed during the time author was a student, so some classmates specified the new number, some—the old one. For the friends, the attribute set contains memberships

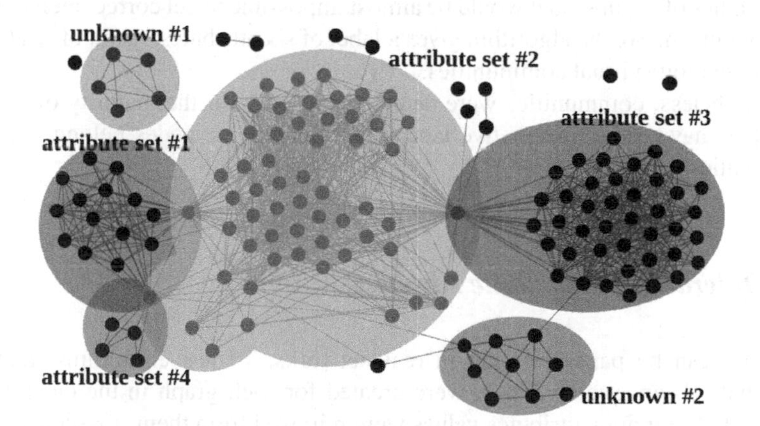

Fig. 2 Communities detected by the developed algorithm for the author's ego-network. Attribute set #1: 4 groups of parties, 4 groups of events, 5 groups of common interests. 3 Attribute set #2: chair, official group of chair, group of each of four groups. Attribute set #3: school number, school old number, official group of school. Attribute set #4: faculty

to several groups by interests like festivals or parties. Neither of these groups has all nodes of community as members, but they give a quite precise idea about author's friends' interests. Author's wife's friends are graduated from the same university and the algorithm gave the corresponding label for them.

The community of pre-university courses acquaintances shows the importance of the last step of the algorithm, where communities of vertices without common attribute are obtained. Indeed, people in this community do not share neither university nor virtual community, so it was detected because it has dense structure.

The community corresponding to workplace detected by the algorithm is missing one node. This is feasible because it is not connected to any other node. This community is also missing a label, but only two of the six vertices has the attribute corresponding to workplace is specified.

The algorithm failed to obtain three small communities. Two of them are trivial and undetectable by the construction. The third one has few edges between nodes and vertices do not share any attribute.

There are more errors in detected university chair community. It has eight extra nodes and university group community was not detected. These eight nodes do have a lot of edges to other nodes in the community, and it is not surprising, because they correspond to wives and girlfriends of people in this community. So instead of "university classmates" community "university classmates and their wifes/girlfriends" community was obtained. Close friends is somehow artificial community—this is just a group of people, who meet each other more often. So, if edge weights corresponding to frequency of meetings would be provided, these nodes could be separated. This may be a topic of further work.

The lack of university group community can also be explained. During the education, there were a lot of transfers between groups, so it was difficult even for the author to separate his group. Also, most of these people were members of all four virtual communities of groups, so it would be almost impossible to get correct memberships even for human. So, the algorithm gives a label of six attributes: chair, official group of chair and four virtual communities.

Nevertheless, communities were detected correctly for the majority of vertices. Also, they have quite informative labels. Note that some nodes belong to several communities.

4.5 *Tolerance to Attribute Absence*

Attributes can be partly missing in real networks. To test algorithms' tolerance to attribute absence, nine copies were created for each graph in the datasets, and $10, 20 \ldots 90\%$ random attributes values were removed from them. F_1-scores by percent of absent attributes for all graphs are shown in Figs. 3, 4 and 5. Graphs for Jaccard index are similar.

F_1-score stays the same for Infomap, modularity maximization, BigCLAM, and AGM-fit algorithms because they use only information about edge structure and do

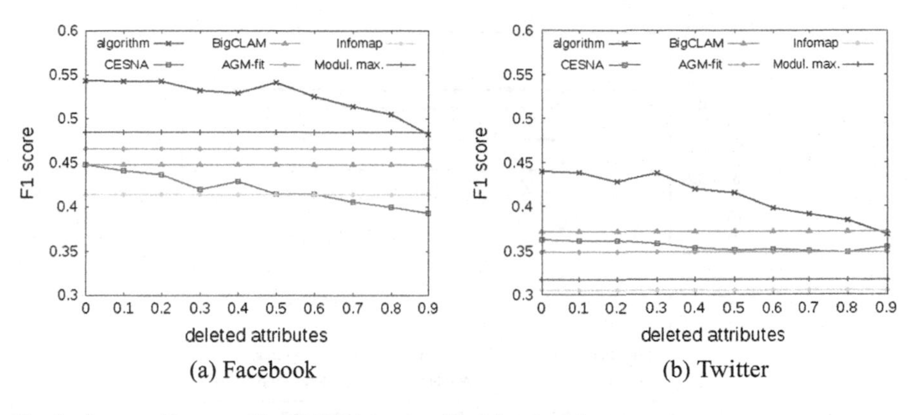

(a) Facebook (b) Twitter

Fig. 3 Average F_1-scores for SNAP datasets with deleted attributes

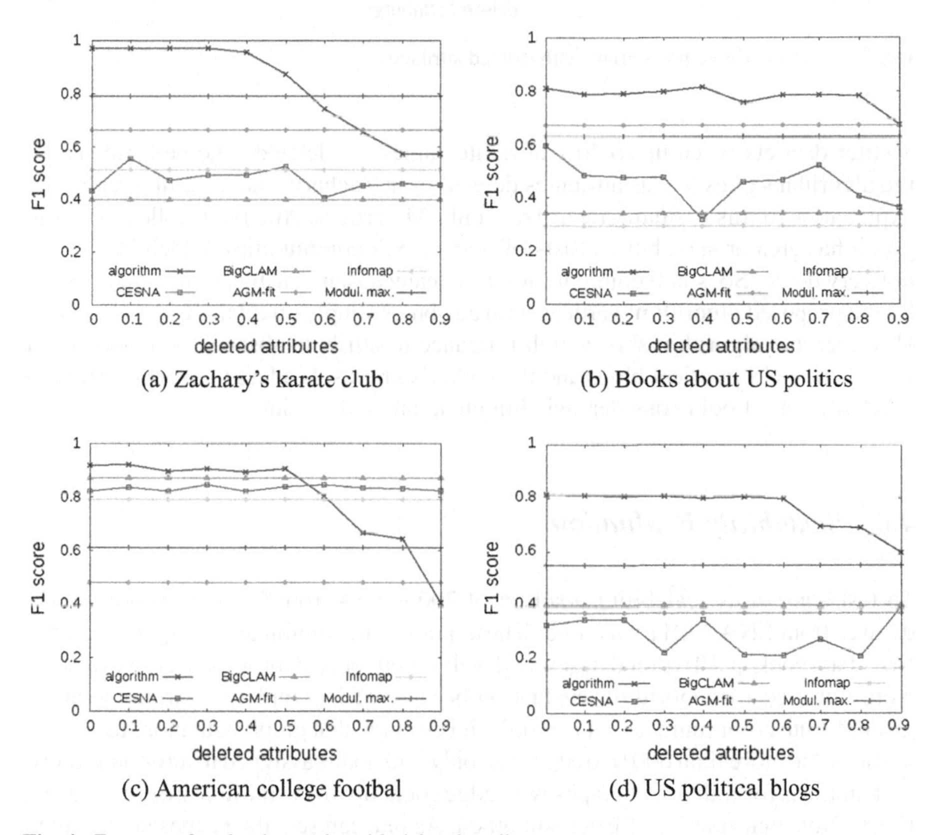

(a) Zachary's karate club (b) Books about US politics

(c) American college footbal (d) US political blogs

Fig. 4 F_1-scores for classic graphs with deleted attributes

not use node attributes at all. One can see that the proposed algorithm show almost the same F_1-score when up to 50% of node attributes values are removed on all graphs. The proposed method still outperforms other algorithms on Facebook and

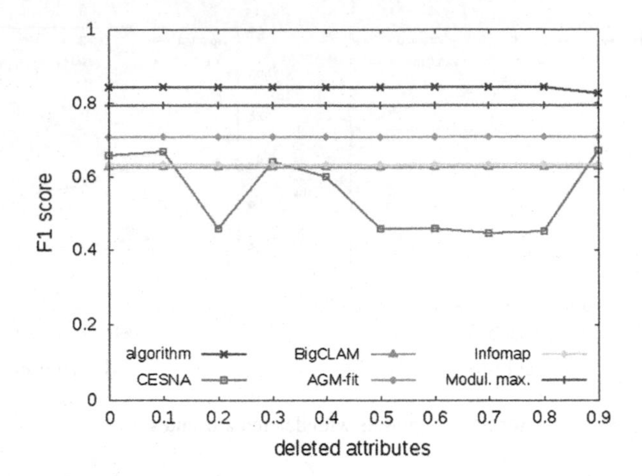

Fig. 5 F_1-scores for author's graph with deleted attributes

Twitter datasets when up to 70% attribute values are deleted. The performance of the algorithms goes low as attributes deleted from Zachary's karate club graph. One explanation to this is small graph size—only 34 vertices. American college football graph has greater size, but consists of many small communities, which in turn are not very dense. So, small communities with many absent attributes are hard to detect by the proposed algorithm: vertices with attributes simply could not form a majority. However, the algorithm shows high tolerance to attribute absence on books about US politics, US political blogs and the author's graphs. If a less than 30% attributes available, one should consider switching to another algorithm.

4.6 Scalability Evaluation

To test algorithms scalability, a subset of 200 graphs from Google+ ego-networks dataset from SNAP [21] was used. These graphs are significantly larger than ones from Facebook and Twitter datasets and well suited for scalability test. However, they were not used for algorithms evaluation because of incomplete information about ground-truth communities. For example, it is strange that graph with more than 4500 vertices and more than 600 000 edges has only 200 nodes assigned to any community.

Runtimes for individual graphs with edge count up to one million are presented on Fig. 6. Note that lines are Bezier smoothed. As one can see, the proposed algorithm is faster than all other contestants except modularity maximization and its runtime grows very slow. AGM-fit algorithm shows the worst performance, while BigCLAM, which was initially developed for community detection in large networks, shows quite good speed.

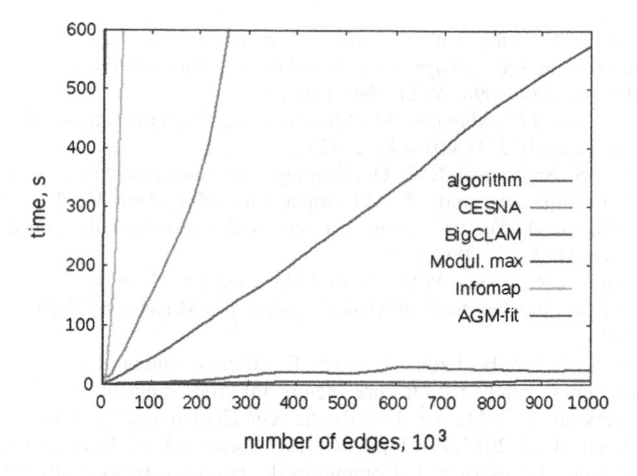

Fig. 6 Algorithms runtime comparison

5 Conclusion

A simple, fast, and scalable algorithm for overlapping community detection has been proposed in the present paper. The developed algorithm uses both edge structure and node attributes. The core idea of the method is attribute transfer from node neighbors. The algorithm is tolerant to the absence of up to half of attributes values. It outperforms Infomap, modularity maximization, BigCLAM, AGM-fit, and CESNA by F_1-score and Jaccard index on Facebook and Twitter datasets, four classic graphs and the author's graph. The algorithm also gives labels to detected communities, showing attributes which formed community.

Further improvement of the algorithm may include method extension to handle weighted or directed graphs. Also, a fast universal heuristic for calculating qualified majority fractions for attributes can be developed.

Acknowledgements This work was supported by RFBR (grant #16-29-09517 "OFI M"). Thanks to Peter Klucharev for critically reading the manuscript and suggesting substantial improvements. The author is also very grateful to the laboratory #11 of ICS RAS staff for productive discussion and critical view.

References

1. Velts, S.V.: Modelling information warfare in social networks based on game theory and dynamic bayesian networks. Eng. J. Sci. Innovation **11**(23) (2013)
2. Leskovec, J., Mcauley, J.J.: Learning to discover social circles in ego networks. In: Pereira, F., Burges, C.J.C., Bottou, L., Weinberger, K.Q. (eds.) Advances in Neural Information Processing Systems 25, pp. 539–547. Curran Associates, Inc. (2012)

3. Ruan, Y., Fuhry, D., Parthasarathy, S.: Efficient community detection in large networks using content and links. In: Proceedings of the 22nd International Conference on World Wide Web, WWW 2013, pp. 1089–1098. ACM, New York (2013)
4. Bothorel, C., Cruz, J.D., Magnani, M., Micenkov, B.: Clustering attributed graphs: models, measures and methods. **3**(3), 408–444 (2015)
5. Xie, J., Kelley, S., Szymanski, B.K.: Overlapping community detection in networks: the state-of-the-art and comparative study. ACM Comput. Surv. 45(4), 43:1–43:35 (2013)
6. Huang, X., Cheng, H., Yu, J.X.: Dense community detection in multi-valued attributed networks. Inf. Sci. **314**(C), 77–99 (2015)
7. Balasubramanyan, R., Cohen, W.W.: Block-LDA: jointly modeling entity-annotated text and entity-entity links. In: Handbook of Mixed Membership Models and Their Applications, pp. 255–273 (2014)
8. Gunnemann, S., Boden, B., Farber, I., Seidl, T.: Efficient mining of combined subspace and subgraph clusters in graphs with feature vectors. In: Proceedings of the Advances in Knowledge Discovery and Data Mining: 17th Pacific-Asia Conference, PAKDD 2013, Gold Coast, Australia, April 14–17, 2013, Part I, pp. 261–275. Springer, Heidelberg (2013)
9. Yang, J., McAuley, J., Leskovec, J.: Community detection in networks with node attributes. In: 2013 IEEE 13th International Conference on Data Mining, Dallas, TX, USA, December 7–10, 2013, pp. 1151–1156 (2013)
10. Nguyen, H.T., Dinh, T.N.: Unveiling the structure of multi-attributed networks via joint non-negative matrix factorization. In: 2015 IEEE Military Communications Conference, MILCOM 2015, pp. 1379–1384 (2015)
11. Hu, L., Chan, K.C.C.: Fuzzy clustering in a complex network based on content relevance and link structures. IEEE Trans. Fuzzy Syst. **24**(2), 456–470 (2016)
12. McPherson, M., Lovin, L.S., Cook, J.M.: Birds of a feather: homophily in social Networks. Ann. Rev. Soc. **27**(1), 415–444 (2001)
13. Bhattacharyya, P., Garg, A., Shyhtsun Felix, W.: Analysis of user keyword similarity in online social networks. Soc. Netw. Anal. Min. **1**(3), 143–158 (2010)
14. Davis Jr., C.A., Pappa, G.L., de Oliveira, D.R.R., de L Arcanjo, F.: Inferring the location of twitter messages based on user relationships. Trans. GIS **15**(6), 735–751 (2011)
15. Mislove, A., Viswanath, B., Gummadi, K.P., Druschel, P.: You are who you know: Inferring user profiles in online social networks. In: Proceedings of the Third ACM International Conference on Web Search and Data Mining, WSDM 2010, pp. 251–260. ACM, New York (2010)
16. Yapan Dougnon, R., Fournier-Viger, P., Nkambou, R.: Inferring user profiles in online social networks using a partial social graph. In: Proceedings of the Advances in Artificial Intelligence: 28th Canadian Conference on Artificial Intelligence, Canadian AI 2015, Halifax, Nova Scotia, Canada, June 2-5, 2015, pp. 84–99. Springer International Publishing, Cham (2015)
17. Chesnokov, V.O., Klucharev, P.G.: Social graph community differentiated by node features with partly missing information. Sci. Educ. Bauman MSTU **9**, 188–199 (2015)
18. Granovetter, M.: The strength of weak ties. Am. J. Soc. **78**(6) (1973)
19. Yang, J., Leskovec, J.: Overlapping community detection at scale: a nonnegative matrix factorization approach. In: Proceedings of the Sixth ACM International Conference on Web Search and Data Mining, WSDM 2013, pp. 587–596. ACM, New York (2013)
20. Yang, J., Leskovec, J.: Community-affiliation graph model for overlapping network community detection. In: 12th IEEE International Conference on Data Mining, ICDM 2012, Brussels, Belgium, December 10–13, 2012, pp. 1170–1175 (2012)
21. Leskovec, J., Krevl, A.: SNAP Datasets: stanford large network dataset collection (2014). http://snap.stanford.edu/data
22. Zachary, W.: An information flow model for conflict and fission in small groups. J. Anthropol. Res. **33**, 452–473 (1977)
23. Krebs, V.: unpublished
24. Girvan, M., Newman, M.E.J.: Community structure in social and biological networks. Proc. Nat. Acad. Sci. USA, **99**(12), 7821–7826 (2002)

25. Adamic, L.A., Glance, N.: The political blogosphere and the 2004 U.S. election: divided they blog. In: Proceedings of the 3rd International Workshop on Link Discovery, LinkKDD 2005, pp. 36–43. ACM, New York (2005)
26. Rosvall, M., Bergstrom, C.T.: Maps of random walks on complex networks reveal community structure. Proc. Natl. Acad. Sci. **105**(4), 1118–1123 (2008)
27. Clauset, A., Newman, M.E., Moore, C.: Finding community structure in very large networks. Phys. Rev. E **70**, 1–6 (2004)
28. Leskovec, J., Sosič, R.: SNAP: a general purpose network analysis and graph mining library in C++, June 2014. http://snap.stanford.edu/snap
29. Ahn, Y.Y., Bagrow, J.P., Lehmann, S.: Link communities reveal multiscale complexity in networks. Nature **466**(7307), 761–764 (2010)
30. Palla, G., Dernyi, I., Farkas, I., Vicsek, T.: Uncovering the overlapping community structure of complex networks in nature and society. Nature **435**(7043), 814–818 (2005)
31. Airoldi, E.M., Blei, D.M., Fienberg, S.E., Xing, E.P.: Mixed membership stochastic blockmodels. J. Mach. Learn. Res. **9**, 1981–2014 (2008)
32. Coscia, M., Rossetti, G., Giannotti, F., Pedreschi, D.: Demon: A local-first discovery method for overlapping communities. In: Proceedings of the 18th ACM SIGKDD International Conference on Knowledge Discovery and Data Mining, KDD 2012, pp. 615–623. ACM, New York (2012)
33. Streich, A.P., Frank, M., Basin, D., Buhmann, J.M.: Multi-assignment clustering for boolean data. In: Proceedings of the 26th Annual International Conference on Machine Learning, ICML 2009, pp. 969–976. ACM, New York (2009)

Testing Hypothesis on Degree Distribution in the Market Graph

P.A. Koldanov and J.D. Larushina

Abstract In this chapter, problems of testing hypotheses on degree distribution in the market graph and of identifying power law in data are discussed. Research methodology of power law hypothesis testing is presented. This methodology is applied to testing hypotheses on degree distribution in the market graphs for different stock markets. Obtained results are discussed.

1 Introduction

Construction and analyzing of models of the stock market attracted great attention in mathematical modelling since the publication of Portfolio Selection by Markowitz [12], in which the mathematical model for the formation of an optimal portfolio was first proposed. Financial market can be represented as a network in which nodes stand for assets and the edges connecting nodes represent the correlations between returns. First networks based on stock returns correlations were investigated by Mantegna [11], were the *minimum spanning tree* (MST) structure was considered. Different properties of MST were studied by several researches [9, 14, 18, 19].

A new network structure, *market graph*, was introduced and investigated in [2, 3]. This approach has been successfully applied for the analysis of variety of stock markets, for instance the US [3], Iranian [13], Swedish [10], Russian [20] and Chinese [9] markets.

There are various properties of the stock market that can be studied after construction of the network: the distribution of correlations of returns, the minimum spanning tree, the density of edges, the maximum cliques and the maximum independent sets in the market graph. In the present paper, hypotheses on vertex degree

P.A. Koldanov (✉) · J.D. Larushina
National Research University Higher School of Economics, Laboratory of Algorithms and Technologies for Network Analysis, Nizhny Novgorod, Russia
e-mail: pkoldanov@hse.ru

J.D. Larushina
e-mail: larka_2010@mail.ru

V.A. Kalyagin et al. (eds.), *Models, Algorithms, and Technologies for Network Analysis*, Springer Proceedings in Mathematics & Statistics 197,
DOI 10.1007/978-3-319-56829-4_15

distribution in the market graph over different periods of observations is tested. Vertex degree distribution describes the topology of the market graph and it can be useful for understanding the stock market as a complex system.

2 Market Model and Problem Statement

Let $G = (V, E)$ be a simple undirected graph, where $V = 1, \ldots, N$ is the set of vertices, and E is the set of edges. Two distinct vertices u and v are called adjacent if they are connected by an edge. An important characteristic of the graph is the degree distribution of its vertices, where the degree of the vertex is the number of edges incident to this vertex. Weight of the edge is the value assigned to the given edge [8].

Market graph is constructed from the market network in the following way [2]: edge (i, j) is included in the market graph if the correlation between returns of the stocks i and j exceeds a given threshold. It is claimed that vertex degree distribution for many graphs arising in diverse areas follow the power law model [3], so that probability for a random vertex to have degree k is $P(k) \propto k^{(-\alpha)}$. This hypothesis has been considered in a variety of works devoted to the study of the market graph [3, 4, 9].

There are different approaches how to test the power law hypothesis. As it was pointed out in [6], the main drawback of the existing researches in this area is the fact that the usual tests are based on the assumption that the parameters of the power distribution can be estimated by constructing a linear regression in the log scale coordinates. If the resulting distribution approximately falls on a straight line, then power law hypotheses are asserted. However it is important to have a more reliable methodology for testing the power law hypothesis. One such methodology was suggested in [6]. We use it to test the power law hypothesis for degree distribution in the market graph.

3 Methodology for Testing the Power Law Distribution

In this Section, we describe the methodology for testing power law hypothesis proposed in Clauset et al. [6]. The procedure consists of estimating parameters (scaling parameter α and lower bound x_{min}) by maximum likelihood, and then calculating the statistical criteria for rejection hypothesis and, if relevant, rejection alternatives. As we have discrete distribution, all subsequent data manipulation will be described on this basis.

Power law for discrete case takes the following form:

$$p(x) = Pr(X = x) = Cx^{-\alpha} \tag{1}$$

where C is a normalizing constant.

Equation 1 diverges at zero [6], so there must be a lower bound $x \geq x_{min} > 0$. Calculation the normalizing constant with condition $\sum_{x=x_{min}}^{\infty} Cf(x) = 1$ we can rewrite the form of discrete power law as:

$$p(x) = \frac{x^{-\alpha}}{\zeta(\alpha, x_{min})} \quad (2)$$

where $C = \frac{1}{\zeta(\alpha, x_{min})}$ and $\zeta(\alpha, x_{min}) = \sum_{n=0}^{\infty}(n + x_{min})^{-\alpha}$.

The method from [6] can be represented as follows:

1. Estimate the parameters x_{min} and α of the power-law model.

   1.1 Assuming that x_{min} is already known, maximum likelihood estimation of α for discrete case takes the form of:

$$\frac{\zeta'(\hat{\alpha}, x_{min})}{\zeta(\hat{\alpha}, x_{min})} = -\frac{1}{n}\sum_{i=1}^{n} \ln x_i \quad (3)$$

   as it was shown in [1, 5], it can be approximated as:

$$\hat{\alpha} \simeq 1 + n\left[\sum_{i=1}^{n} \ln \frac{x_i}{x_{min} - 1/2}\right]^{-1} \quad (4)$$

   The details of approximation can be found in Appendix B of the paper [6].

   1.2 As it is shown in [16] and references therein, parameter estimation techniques can be sensitive to noise or fluctuations in the tail of distribution, so methodology discussed in [6] involves usage of a method proposed in [5]. General idea is briefly represented below.

   - For all possible x_{min} chosen distance (Kolmogorov–Smirnov statistic [7]) is measured:

$$\mathcal{D} = max_{x \geq x_{min}} |S(x) - P(x)|, \quad (5)$$

   where S(x) - CDF for empirical data, P(x) - CDF for model.

   - Chosen for lower bound $\hat{x}_{min}$ minimizes $\mathcal{D}$

   As a result of this selection, for chosen $\hat{x}_{min}$ empirical distribution is close to best fit model distribution for all $x \geq \hat{x}_{min}$.

   Also this distance can be reweighted [7] in order to solve the problem with insensitivity of Kolmogorov–Smirnov statistic to differences of the distributions at the extreme limits (at the extreme limits the CDFs necessarily tend to zero and one [6]).

$$D^* = max_{x \geq x_{min}} \frac{|S(x) - P(x)|}{\sqrt{P(x)(1 - P(x))}} \quad (6)$$

2. Calculate the goodness-of-fit between the data and the power law.
   This part includes the following steps:

   2.1 Calculation of chosen goodness-of-fit between the data and the best fit model.
   2.2 Generation of M sets of synthetic data from best fit power-law model. M depends on preferable precision of $p - value$. It was claimed [6] that $\frac{1}{4}\varepsilon^{-2}$ synthetic sets are needed for $p - value$ to be accurate to within about ε of the true value.
   2.3 Estimation of parameters for each set (building a best fit model) and calulation of goodness-of-fit between the synthetic data and corresponding best fit model.
   2.4 Calculation of $p - value$ as the fraction of the synthetic distances that are larger than the empirical distance.

   If the resulting value is larger than predefined threshold (in [6] $p - value > 0.1$) the hypotheses cannot be ruled out.
3. Compare the power law with alternative hypotheses via a likelihood ratio test. If power law distribution hypotheses cannot be ruled out after the test, comparison with alternative hypotheses is highly recommended. For two candidate distributions for empirical data set with probability density functions $p_1(x)$ and $p_1(x)$ log likelihood ratio takes the form of:

$$\mathscr{R} = \sum_{i=1}^{n} \left[\ln p_1(x_i) - \ln p_2(x_i)\right] = \sum_{i=1}^{n} \left[l_i^{(1)} - l_i^{(2)}\right]. \tag{7}$$

   Then, sign of $\mathscr{R}$ points at a preferable alternative.

   To determine the statistical significance of the observed sign, there is a method proposed by Vuong [21] and discussed in [6]. This method allows to give numerical estimation on the degree of confidence in the plausibility of tested hypotheses compared with the alternative in contradistinction to competing methods [15, 17], which give no estimation of whether the observed results could be obtained due to chance.

4 Experimental Study of Power Law Distribution for Different Market Graphs

In order to test hypotheses on power law degree distribution of the market graph we put it in the framework of procedure discussed in the previous section.

We collected data on logarithmic stock returns from markets of China, France, Germany, India, Russia, Britain and the United States of America with the following characteristics:

- observation horizon: 1 year (from January 1 to December 31),
- observation periods: 12 periods (2003–2014),
- assets: 100 assets, top by liquidity for the current observation period,
- asset attribute: logarithmic stock returns.

Based on collected data we constructed Pearson correlations network, in which sample correlation between the stocks i and j is defined by

$$\hat{\gamma}^P_{i,j} = \frac{\sum_{t=1}^{n}(x_i(t) - \bar{x_i})(x_j(t) - \bar{x_j})}{\sqrt{\sum_{t=1}^{n}(x_i(t) - \bar{x_i})^2}\sqrt{\sum_{t=1}^{n}(x_j(t) - \bar{x_j})^2}}, \tag{8}$$

where n is number of observations and $x_i(t)$ is logarithmic stock return of asset i in day t. Degree distribution was calculated for a plurality of cut-off thresholds $\theta = \{0.1, \ldots, 0.9\}$ for absolute and initial values of correlation ($\left|\hat{\gamma}^P_{i,j}\right| \geq \theta$ and $\hat{\gamma}^P_{i,j} \geq \theta$). This procedure resulted in 1512 vectors of observations, each of which represents set of degrees of vertices in the corresponding graph, for 1456 of which discussed hypotheses has been tested. The inability to test hypotheses for the remaining 56 data vectors attributed to their length.

We assume each vector element to be a realization of the corresponding random variable, so that we have 1512 random variables and their realisations.

The problem of power law identification of degree distribution in a graph can be formulated as follows:

$$H_0 : F_n(x) \in \{F(x, \theta), \theta \in \Theta\} \tag{9}$$

$$H_1 : F_n(x) \notin \{F(x, \theta), \theta \in \Theta\} \tag{10}$$

where $F(x, \theta) = \frac{x^{-\alpha}}{\zeta(\alpha, x_{min})}, \theta = (\alpha, x_{min})$.
A statistical procedure for hypotheses testing of H_0 can be written in the following form:

$$\varphi(x) = \begin{cases} 1, & T(x) > 0.1 \\ 0, & T(x) \leq 0.1 \end{cases} \tag{11}$$

where $T(x) = \frac{\sum_{i=1}^{M}[D_i > D_{emp}]}{M}$ is the fraction of the Kolmogoro–Smirnov distances from synthetic data to their best-fit models that are larger than the empirical distance and M is the number of synthetic samples. Threshold 0.1 was used taking into account the assumptions in the original research [6]. It is shown there that more lenient rule ($T(x) \leq 0.05$) would let through some candidate distributions that have only a very small chance of really following a power law. To define significance level, it is necessary to find distribution of the test statistic $T(x)$ from Eq. 11. Otherwise significance level can be found by simulations. Procedure for its estimation is going to be thoroughly developed in further studies.

Table 1 Obtained results: dataset

Country	Year	Threshold	abs	alpha	xmin	p_value	Comparison
China	2012	0.60	Initial value	5.86	23.00	0.82	−0.75
Germany	2013	0.40	Absolute value	4.44	15.00	0.39	−0.62
China	2013	0.60	Initial value	8.51	20.00	0.34	−0.58
China	2010	0.10	Absolute value	8.39	69.00	0.56	−0.58
UK	2010	0.80	Absolute value	2.87	27.00	0.00	−0.68

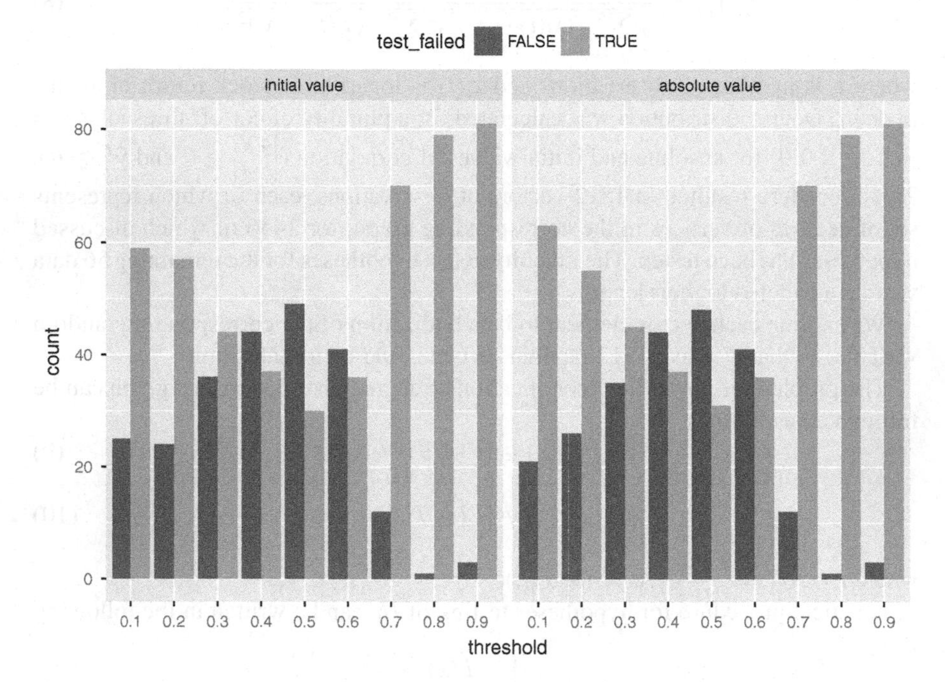

Fig. 1 Hypotheses rejection with respect to cut-off threshold

For each data vector parameter estimation technique and all steps discussed were applied, after which the procedure of testing of statistical hypotheses on the distribution of empirical data has been conducted. Log-normal distribution was chosen as an alternative hypothesis. Dataset structure can be found in Table 1.

Primarily, the cut-off wherein the hypothesis was not rejected in most cases threshold was found. Figure 1 represents frequency of hypotheses rejection with respect to threshold for both initial and absolute values of correlation.

Second, for all cases, in which statistically significant differences frequently were not obtained ($0.3 \leq \theta \leq 0.6$), we have tried to identify a continuous time period for which the assertion of the power law would be true almost always. Figure 2 shows that such a period does not exist.

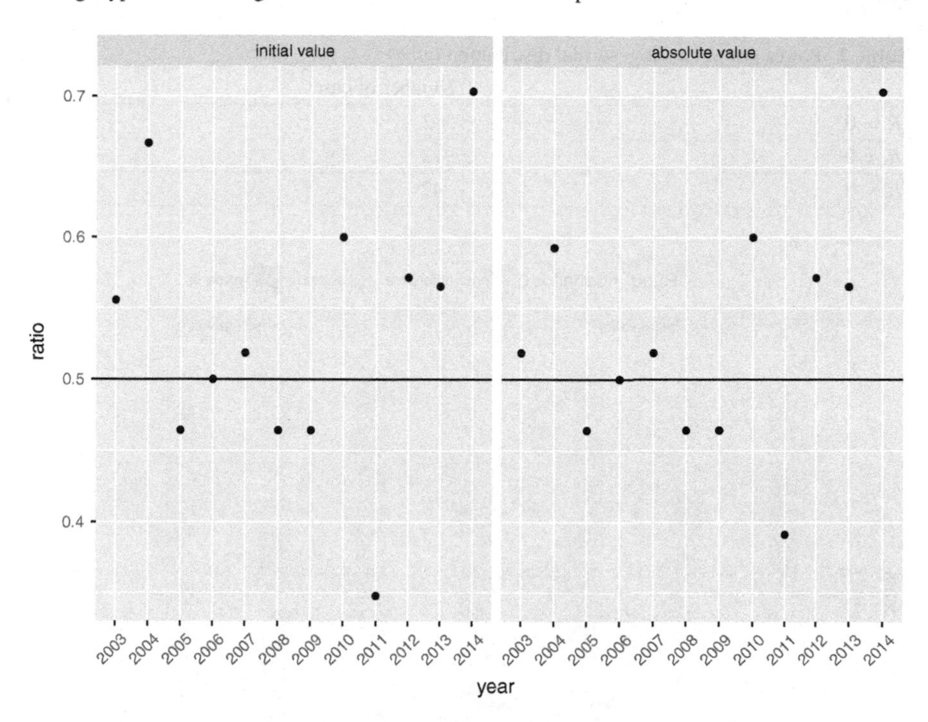

Fig. 2 Hypotheses assertance ratio with respect to year ($0.3 \leq \theta \leq 0.6$)

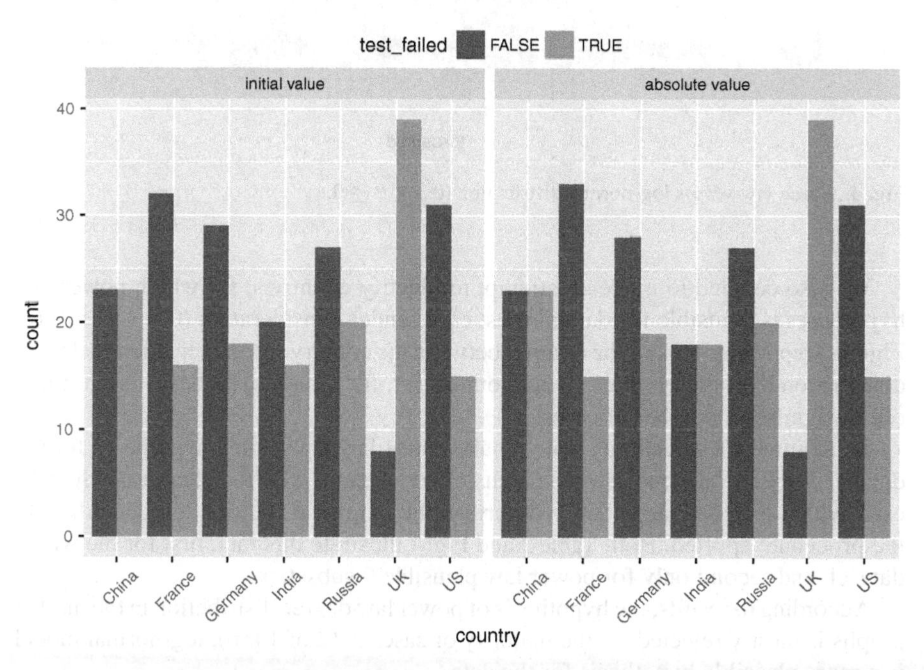

Fig. 3 Hypotheses rejection frequency with respect to country

Table 2 Power law versus log-normal distribution (all θ)

	Number of cases
$R = 0$	502
$R < 0$	912
$R > 0$	42

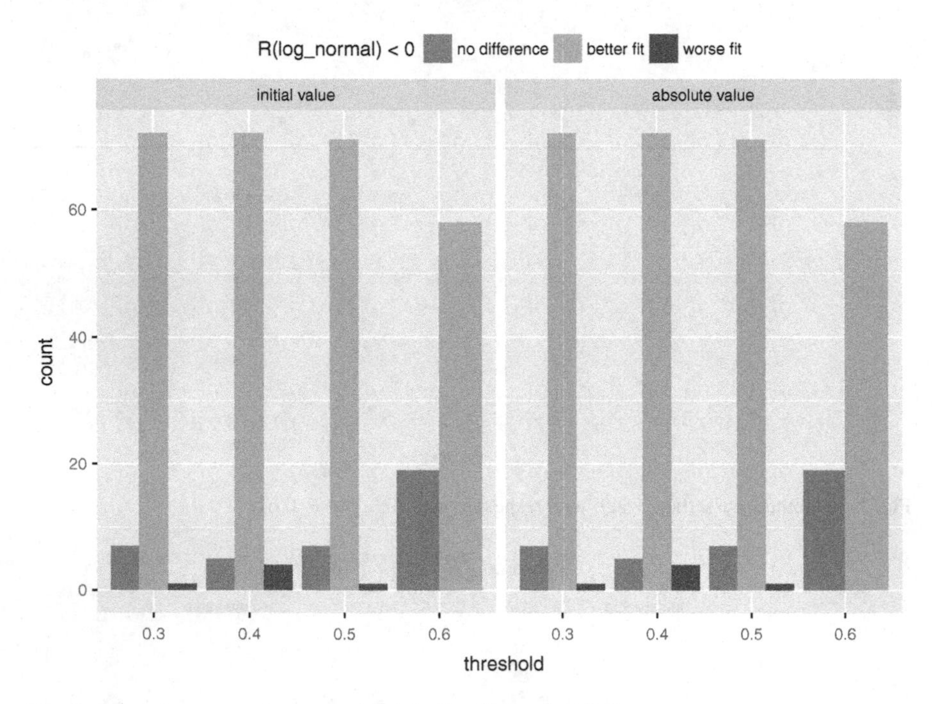

Fig. 4 Power law versus log-normal distribution ($0.3 \leq \theta \leq 0.6$)

We also decided to make an attempt to identify countries, for which power law hypotheses is plausible for data in most cases under consideration ($0.3 \leq \theta \leq 0.6$). Figure 3 represents lack of the relation between the country set and rejection of degree distribution hypotheses: there is no country set, for which hypotheses is not ruled out in significant number of cases.

In addition, for thresholds which make power law a plausible hypotheses for the data ($0.3 \leq \theta \leq 0.6$) one can see results of comparison with the alternative hypothesis, which in most cases better describes the empirical distribution according to the procedure applied. Both Table 2 and Fig. 4 illustrate this fact, first for the whole dataset, and second only for power law plausible θ subset.

According to results, the hypothesis of power law degree distribution in the market graphs is mostly rejected. In the majority of cases (912 of 1456), log-normal model is a more plausible hypothesis for the data.

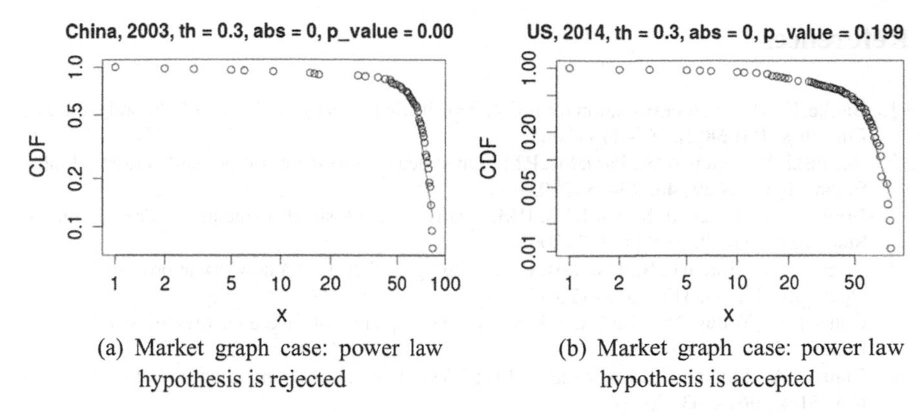

(a) Market graph case: power law hypothesis is rejected

(b) Market graph case: power law hypothesis is accepted

Fig. 5 Instability of visual evaluation

Note, that it is possible to compare described method with visual evaluation. To do this we apply the described method for two cases. First stands for the case, when power law hypothesis is rejected, second stands for the case where it is accepted. The Fig. 5 shows that visual evaluation cannot distinguish these two cases.

5 Concluding Remarks

In the present paper, power law hypotheses of degree distribution for the market graph are tested and alternative hypotheses are proposed. Our main findings are:

- Current approaches of identifying power law in the market graph can cause significant errors up to detection of the power law in cases when the plausibility of other distributions of much higher.
- There is a cut-off threshold for Pearson correlation network, which results in high probability of power law ($0.3 \leq \theta \leq 0.6$) in comparison with other thresholds.
- Proposed statistical procedure applied to the stock markets leads to rejection of power law distribution of vertex degrees in the market graph for significant majority of cases.
- Numerical experiments show that it is meaningful to test the hypothesis on lognormal degree distribution of the market graph.

Acknowledgements The work of Koldanov P.A. was conducted at the Laboratory of Algorithms and Technologies for Network Analysis of National Research University Higher School of Economics. The work is partially supported by RFHR grant 15-32-01052.

References

1. Bauke, H.: Parameter estimation for power-law distributions by maximum likelihood methods. Eur. Phys. J. B **58**(2), 167–173 (2007)
2. Boginski, V., Butenko, S., Pardalos, P.M.: On structural properties of the market graph. Innov. Financ. Econ. Netw. **48**, 29–45 (2003)
3. Boginski, V., Butenko, S., Pardalos, P.M.: Statistical analysis of financial networks. Comput. Stat. Data Anal. **48**(2), 431–443 (2005)
4. Boginski, V., Butenko, S., Pardalos, P.M.: Mining market data: a network approach. Comput. Oper. Res. **33**(11), 3171–3184 (2006)
5. Clauset, A., Young, M., Gleditsch, K.S.: On the frequency of severe terrorist events. J. Confl. Resolut. **51**(1), 58–87 (2007)
6. Clauset, A., Shalizi, C.R., Newman, M.E.: Power-law distributions in empirical data. SIAM Rev. **51**(4), 661–703 (2009)
7. Flannery, B.P., Press, W.H., Teukolsky, S.A., Vetterling, W.: Numerical recipes in c, p. 24. Press Syndicate of the University of Cambridge, New York (1992)
8. Harary, F., et al.: Graph theory (1969)
9. Huang, W.Q., Zhuang, X.T., Yao, S.: A network analysis of the chinese stock market. Phys. A Stat. Mech. Appl. **388**(14), 2956–2964 (2009)
10. Jallo, D., Budai, D., Boginski, V., Goldengorin, B., Pardalos, P.M.: Network-based representation of stock market dynamics: an application to american and swedish stock markets. In: Models, Algorithms, and Technologies for Network Analysis, pp. 93–106. Springer (2013)
11. Mantegna, R.N.: Hierarchical structure in financial markets. Eur. Phys. J. B Condens. Matter Complex Syst. **11**(1), 193–197 (1999)
12. Markowitz, H.M.: Portfolio Selection: Efficient Diversification Of Investments, vol. 16. Yale university press, New Haven (1968)
13. Namaki, A., Shirazi, A., Raei, R., Jafari, G.: Network analysis of a financial market based on genuine correlation and threshold method. Phys. A Stat. Mech. Appl. **390**(21), 3835–3841 (2011)
14. Onnela, J.P., Kaski, K., Kertész, J.: Clustering and information in correlation based financial networks. Eur. Phys. J. B Condens. Matter Complex Syst. **38**(2), 353–362 (2004)
15. Shalizi, C.R., et al.: Dynamics of bayesian updating with dependent data and misspecified models. Electron. J. Stat. **3**, 1039–1074 (2009)
16. Stoev, S.A., Michailidis, G., Taqqu, M.S.: Estimating heavy-tail exponents through max self-similarity. arxiv preprint arXiv:math/0609163 (2006)
17. Stouffer, D.B., Malmgren, R.D., Amaral, L.A.: Comment on barabási. Nature **435**, 207 (2005)
18. Tumminello, M., Di Matteo, T., Aste, T., Mantegna, R.: Correlation based networks of equity returns sampled at different time horizons. Eur. Phys. J. B **55**(2), 209–217 (2007)
19. Tumminello, M., Lillo, F., Mantegna, R.N.: Correlation, hierarchies, and networks in financial markets. J. Econ. Behav. Organ. **75**(1), 40–58 (2010)
20. Vizgunov, A., Goldengorin, B., Kalyagin, V., Koldanov, A., Koldanov, P., Pardalos, P.: Network approach for the russian stock market. Comput. Manag. Sci. **11**(1–2), 45–55 (2014)
21. Vuong, Q.H.: Likelihood ratio tests for model selection and non-nested hypotheses. Econom. J. Econom. Soc. 307–333 (1989)

Application of Network Analysis for FMCG Distribution Channels

Nadezda Kolesnik, Valentina Kuskova and Olga Tretyak

Abstract The paper presents the approach for multidimensional analysis of marketing tactics of the companies employing network tools. The research suggests omni-channel distribution tactic of a company as a node in eight-dimensional space. Dimensions for node location are defined by frequency of usage of eight communication channels (friends, acquaintances, telephone, home presentations, printed advertisement, internet, e-mail, and door to door). The comparison is grounded on measuring pairwise distance between nodes in eight-dimensional space. Pairwise distance measured by Euclidean norm is used as a weight of edge between companies. The smaller the Euclidean distance, the higher is similarity. Further, we employ network representation of multidimensional statistics to analyze performance and companies' characteristic, such as product category, market share, education level, and average age of distributors. Empirical implication is approved on the sample from 5694 distributors from 16 fast moving consumer goods (FMCG) distributing companies from direct selling industry.

1 Introduction

In direct selling traditionally communication with customers is the key aspect for distribution channel successes. Communication is based on demonstrations and personal engagement that make the buying process highly tangible and multisensory for consumers [7]. The close involvement with the product and personal communication has been a key differentiating feature of direct selling when compared to

N. Kolesnik (✉) · V. Kuskova · O. Tretyak
National Research University Higher School of Economics, 20 Myasnitskaya Ulitsa, Moscow, Russia
e-mail: nadezda.kolesnik@mail.ru

V. Kuskova
e-mail: vkuskova@hse.ru

O. Tretyak
e-mail: o_tretyak@inbox.ru

V.A. Kalyagin et al. (eds.), *Models, Algorithms, and Technologies for Network Analysis*, Springer Proceedings in Mathematics & Statistics 197,
DOI 10.1007/978-3-319-56829-4_16

other methods of sales and communication with potential customers [11]. Usage of internet technologies for communication may improve productivity, but challenge the customary "high-touch" tradition in the industry [4].

Previously, studies have mainly considered separately offline and online channels [5, 6]. In our research, we estimate usage of all channels simultaneously employing omni-channel concept. Omni-channel retailing reflects the fact that salesperson interacts with customers through numerous channels. Integration allows taking advantages of digital channels (wide selection, rich information, and reviews) and advantages of physical stores (personal service, the ability to touch products, convenient returns, and shopping experience) [12]. In omni-channel concept, different channels become blurred as the natural borders between channels begin to disappear [15]. Our research is designed to compare multidimensional tactics as a combination of different communication channels for distribution. Each combination of channels is considered an omni-channel tactic of a company (friends, acquaintances, telephone, home presentations, printed advertisement, internet, e-mail, and door to door). The purpose of this study is to investigate the distribution tactics of different small-, medium-, and big-sized FMCG companies in DS industry in relation to performance.

2 Direct Selling Distribution Channel

The paper is illustrated with data from direct selling (DS) industry. Direct Selling Associations define DS as a method of marketing and retailing goods and services directly to the consumers, in any location away from permanent retail premises. The company provides partners with the opportunity to build own business with full- or part-time employment and getting profit from it. For consumers, it is an alternative to stationary shops and allows to make purchases in a convenient location with a personal approach.

According to World Federation Direct Selling Association (WFDSA) report, worldwide sales at direct selling companies were 182.8 billion US dollars in 2014, with a sales force estimated at 99,7 million independent contractors worldwide [16]. Sales in the Europe countries in 2014 were reported to be 32,6 billion and the number of sales people was 13,97 million.

Russian Federation is the world's 11th largest direct selling market. In 2014, direct retail sales were in excess of 3.6 billion US dollars, showing a 3-year compound annual growth rate of 1.6% (2011–2014). Number of independent direct sellers showed a 7.6% increase to 5.4 million in 2014 (5 425 830).

On emerging markets, direct selling industry fills the gap in weak distribution system, especially in regions. For citizens, direct selling is the source for additional income. Distributors are considered as individual small enterprises [1].

In the paper, we employ the term distributor based on the functions implemented. DS distributor is an independent contractor who performs functions of retailers—promote, sell, and distribute products and services to consumers [11]. An integral aspect of the DS industry is a personal presentation in a face-to-face manner [13].

Products are demonstrated to an individual, or to a group or where a catalogue is left with the consumer and where the direct seller calls later to collect orders.

3 Sample Description

Data for the research was collected jointly with the Russian Direct Selling Association (RDSA), (www.rdsa.ru) member of World Federation Direct Selling Association in spring 2014. The questionnaires were spread among companies with RDSA membership, the sample was quoted according to the number of the distributors. The sampling frame included 5694 independent sellers from 16 biggest direct selling companies in Russia. The companies are focused on distribution of FMCG: personal care products, nutrition, perfumes, cosmetics, hygiene products, jewelry, and accessories. Table 1 summarizes the descriptive statistics of the sample.

In the research, we consider only single-level direct selling companies, where salespeople devote all efforts to selling and achieve all compensation based on their own sales and do not build an organization via recruiting and training [2]. The sampling frame includes sellers from 16 DS companies: Amway, AVON, Faberlic, Florange, Herbalife, LR Health & Beauty Systems, Mary Kay, Tapperware, Mirra, Nu Skin, Oriflame, Nikken, Jafra, CIEL, Tentorium, Morinda. The companies are focused on distribution of products for personal care, nutrition, beauty, and household: perfumes, cosmetics, hygiene products, jewelry, and accessories.

4 Channels and Tactics

Companies use variety of channels for communication with customers and distribution. In the research, we analyze eight communication channels used by direct sellers. Table 2 presents statistics of channel usage frequency. Sales via personal communication with friends and acquaintance are on the top lines.

According to the definition of Cambridge dictionary, friends are people who are known well and liked a lot. Usually it is a small number of people from close surrounding. Acquaintance is defined as a person that you have met but do not know well. It might be classmates, colleagues, and different group's mates.

Different channels within each tactic can be used by distributors with different frequency. Frequency is relative to the time salesperson spends for selling activities, which varies from 1 hour up to 40 hours per week. Frequency is measured by four-point scale from "never use=1", "rarely=2", "occasionally=3", and "frequently=4". The *Cronbach alpha* value of the scale is equal to "0.77", indicating satisfactory internal reliability. Figure 1 shows frequency of different channels usage for different age groups within the total sample.

Totally there are 256 different combinations from eight individual channels. Each combination is considered as an omni-channel tactic of salesperson. We use

Table 1 Sample descriptive statistics

		Number of salespersons	Percentage (%)
Age groups	<18	28	0.5
	19–24	492	8.7
	25–30	775	13.6
	31–34	502	8.8
	35–40	684	12.0
	41–50	1159	20.4
	51–55	789	13.9
	56–65	973	17.1
	>65	278	4.9
Work experience	Less than one year	1251	22.2
	1 year	543	9.6
	2 years	599	10.6
	3 years	503	8.9
	4 years	373	6.6
	5 years	430	7.6
	6 years	335	5.9
	7–10 years	791	14.0
	More than 10 years	808	14.3
Location size	Over 1 mln. people	2579	46.2
	500 thousand–1 mln. people	957	17.2
	100 thousand–500 thousand people	1017	18.2
	10 thousand–100 thousand people	693	12.4
	Less than 10 thousand people	332	6.0
Hours spent for work, per week	<1	672	12.0
	1–4	1704	30.5
	5–9	1076	19.2
	10–14	656	11.7
	15–19	414	7.4
	20–29	468	8.4
	30–40	301	5.4
	>40	304	5.4

(continued)

Table 1 (continued)

		Number of salespersons	Percentage (%)
Income per month, rub.	<3,000	2002	36.1
	3,000–4,999	825	14.9
	5,000–9,999	731	13.2
	10,000–14,999	540	9.7
	15,000–24,999	492	8.9
	25,000–34,999	276	5.0
	35,000–49,999	241	4.3
	>50,000	434	7.8
Total		5,694	100

Table 2 Breakdown by channel usage (Percentage, amount of salespersons)

Communication channels	Do not use (%)	Rarely (%)	Occasionally (%)	Frequently (%)
Friends	18	7.3	20.7	55.3
Acquaintance	18	7.9	24.3	50.7
Telephone	42	10.3	15.1	33.0
Home presentations	44	16.4	20.5	20.2
Printed advertisement	47	14.5	20.7	18.7
Internet	54	13.0	15.3	18.7
E-mail	66	12.2	11.9	9.9
Door to door	85	9.8	3.4	1.8

two-mode graph to analyze structure and relationships of tactics within the industry. To better understand the structure and effectiveness of the company's distribution network [8, 10], we examine the effects of the network structure on the performance.

Figure 2 depicts the overall two-mode network for 256 tactics used by 5694 independent sellers. It is seen that some tactics are more popular than other within distributors; 66 tactics are not used at all. Out of 190 tactics, only 130 are used by more than one salesperson. Unfortunately, the total picture is not rather clear due to the large amount of distributors and tactics. Therefore, we agglomerate the object of the research. Further, we are focusing on the company's tactic level.

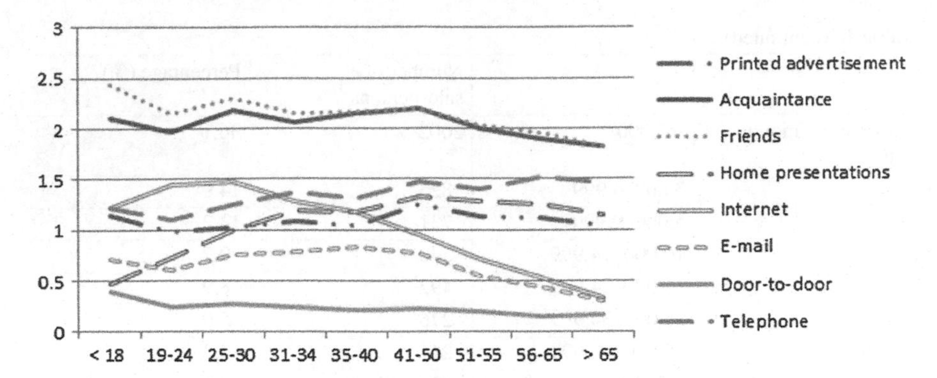

Fig. 1 Distribution channels used by different age groups

Fig. 2 Two-mode network of the sample (distribution tactics and distributors)

5 Comparison of Communication Tactics in Distribution Channels

Distribution tactic of the company is a set of communication channels, which are used with different frequency. Table 3 presents frequency of different channels usage for each company. All companies employ existing channels, but with different frequencies. In our sample, number of respondents for each company in average is

Table 3 Frequency of distribution channel usage

Channels Company	Friends	Acquaintances	Tele phone	Home presen-tations	Printed adver-tise-ment	Internet	E-mail	Door to door
A	0.30	1.90	2.18	1.85	0.52	0.30	0.13	1.30
B	1.23	2.12	1.97	0.70	0.83	0.55	0.18	1.18
C	0.77	2.07	2.13	1.12	1.52	0.87	0.19	1.57
D	1.44	2.57	2.57	1.89	1.80	1.42	0.14	2.08
E	1.67	2.22	2.25	2.06	0.51	0.42	0.14	1.77
F	1.06	1.87	1.80	0.68	0.46	0.27	0.05	0.88
G	0.97	2.12	2.04	1.11	0.88	0.85	0.17	1.71
H	1.22	1.99	2.16	0.62	1.39	0.73	0.29	0.95
I	1.34	1.99	1.95	1.03	1.05	0.44	0.62	1.35
J	0.87	1.89	1.98	0.99	0.50	0.48	0.11	1.66
K	1.12	2.18	2.19	1.77	0.79	0.75	0.31	1.84
L	2.26	2.44	2.40	1.08	1.58	0.79	0.16	1.24
M	1.20	2.29	2.15	1.09	1.02	0.62	0.17	1.41
N	0.88	1.66	1.70	0.54	0.66	0.54	0.15	0.96
O	1.27	1.83	1.89	0.72	0.72	0.61	0.17	1.14
P	1.04	1.89	2.06	0.55	1.06	0.58	0.38	0.84

equal to 350. Frequency of channel usage for each company was calculated as mean observation.

To estimate similarity of the tactics we use Euclidean norm concept [9, 14]. We estimate how far points between each other in eight-dimensional space are. Node location for each company is described by a node in eight-dimensional space. For example, point position for Amway company is derived from frequency of each channel usage and coded as (0.30; 1.90; 2.18; 1.85; 0.52; 0.30; 0.13; 1.30). We calculate pairwise distance between nodes employing Euclidean norm.

On a n-dimensional Euclidean space *Rn*, the intuitive notion of length of the vector $x = (x_1, x_2, ..., x_n)$ is captured by the formula

$$\|\overrightarrow{X}\| = \sqrt{x_1^2 + \cdots + x_n^2}, \tag{1}$$

where $x_i$location of node identified by usage of distribution channel i.

We employ equation for eight-dimensional Euclidean space, as company's tactic is described by eight parameters.

$$\begin{aligned}\|\overrightarrow{X_p}\|_2 &= \sqrt{\|X_1 - X_2\|_2} \\ &= \sqrt{(x_{11} - x_{21})^2 + (x_{12} - x_{22})^2 + \cdots + (x_{18} - x_{28})^2}, \end{aligned} \tag{2}$$

Table 4 Pairwise distance (Euclidean norm)

	A	B	C	D	E	F	G	H	I	J	K	L	M	N	O	P
A	0	1,56	1,48	2,32	1,5	1,51	1,29	1,86	1,53	1,12	1,16	2,48	1,38	1,58	1,57	1,69
B	1,5	0	1,07	2,12	1,61	0,67	0,78	0,69	0,65	0,78	1,3	1,45	0,55	0,71	0,34	0,57
C	1,4	1,07	0	1,46	1,73	1,55	0,69	0,94	1	1,14	1,09	1,6	0,76	1,38	1,19	1,13
D	2,3	2,12	1,46	0	1,75	2,69	1,62	2,03	1,93	2,17	1,4	1,59	1,62	2,65	2,27	2,34
E	1,5	1,61	1,73	1,75	0	1,85	1,33	1,97	1,42	1,41	0,78	1,71	1,26	2,05	1,65	2,01
F	1,5	0,67	1,55	2,69	1,85	0	1,24	1,15	1,08	0,9	1,66	1,98	1,1	0,48	0,56	0,81
G	1,2	0,78	0,69	1,62	1,33	1,24	0	1,09	0,83	0,61	0,73	1,61	0,5	1,16	0,87	1,12
H	1,8	0,69	0,94	2,03	1,97	1,15	1,09	0	0,83	1,3	1,58	1,3	0,83	1,01	0,79	0,45
I	1,5	0,65	1	1,93	1,42	1,08	0,83	0,83	0	0,95	1,09	1,37	0,63	1,07	0,72	0,82
J	1,1	0,78	1,14	2,17	1,41	0,9	0,61	1,3	0,95	0	1,02	1,97	0,82	0,91	0,76	1,13
K	1,1	1,3	1,09	1,4	0,78	1,66	0,73	1,58	1,09	1,02	0	1,7	0,86	1,7	1,37	1,63
L	2,4	1,45	1,6	1,59	1,71	1,98	1,61	1,3	1,37	1,97	1,7	0	1,25	2,06	1,59	1,64
M	1,3	0,55	0,76	1,62	1,26	1,1	0,5	0,83	0,63	0,82	0,86	1,25	0	1,15	0,77	0,92
N	1,5	0,71	1,38	2,65	2,05	0,48	1,16	1,01	1,07	0,91	1,7	2,06	1,15	0	0,53	0,66
O	1,5	0,34	1,19	2,27	1,65	0,56	0,87	0,79	0,72	0,76	1,37	1,59	0,77	0,53	0	0,6
P	1,6	0,57	1,13	2,34	2,01	0,81	1,12	0,45	0,82	1,13	1,63	1,64	0,92	0,66	0,6	0
Total distance	24,03	14,84	18,22	29,96	24,02	19,24	15,48	17,83	15,93	17,00	19,07	25,31	14,41	19,11	15,56	17,54

where $\mathbf{x}_{11}$ is the average meaning for channel 1 of company 1 and $\mathbf{x}_{21}$ average meaning for channel 1 of company 2.

We have introduced parameter "total distance", which is equal to the sum of all distances of the company with other companies. Parameter "total distance" reflects uniqueness of the companies' tactics. Also, it is used for comparison of the companies and further analysis.

We use results from the Table 4 for network analysis. We draw the network, where nodes are companies. Nodes are connected if companies use the same channels in their tactics. On the company level, all companies use all channels, therefore we have complete graph. Weight of the edge on the graph is measured by pairwise distance. Smaller distance designates closer connection between companies.

The evaluation determined by the Euclidean norm makes us understand the similarity by using an expression based on the concept of norm. The smaller the Euclidean, the higher is similarity. Similarity of the tactics is based on the simultaneous evaluation of usage frequency of all eight communication channels. Further, we employ network visualization tools to analyze performance and companies' characteristic, such as product category, market share, education level, and average age of distributors.

Table 5 Direct selling by category: % Value Growth 2009–2014

% current value growth, retail value excl. sales tax	2013/14	2009-14 CAGR	2009/14
Apparel and footwear	3.9	38.0	401.2
Beauty and personal care	−3.2	−0.8	−3.9
Consumer healthcare	15.5	27.1	231.7
Consumer appliances	8.9	13.7	90.1
Home care	6.3	8.7	51.7
Home improvement and gardening	25.7	10.1	62.0
Housewares and home furnishings	−1.7	−3.9	−18.1
Direct selling	1.5	4.9	27.3

Source Euromonitor International from official statistics, trade associations, trade press, company research, trade interviews, trade sources

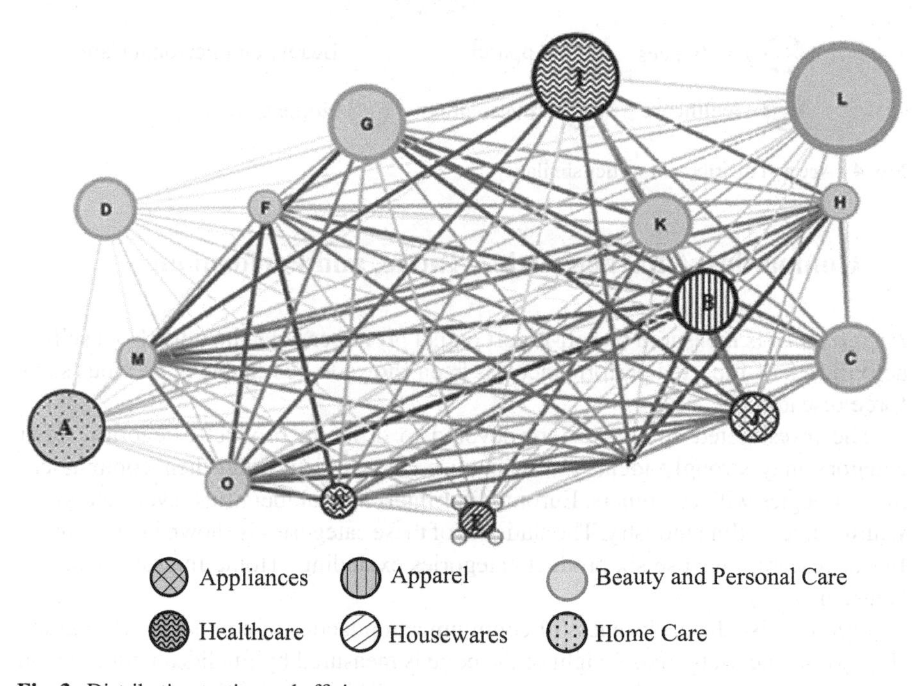

Fig. 3 Distribution tactics and efficiency

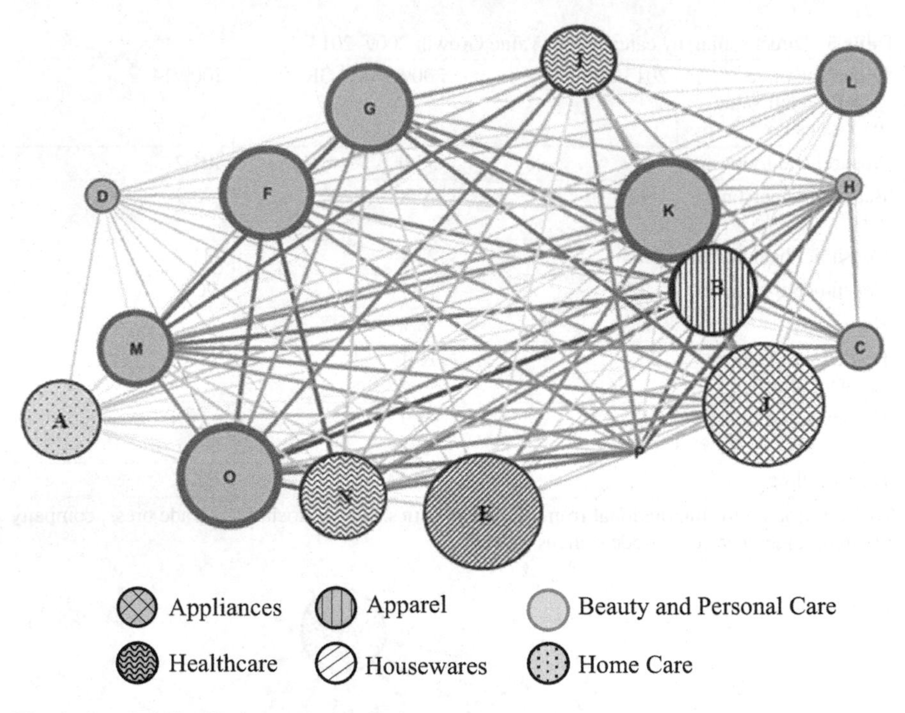

Fig. 4 Age and distribution tactics similarity

6 Companies' Characteristics, Tactic, and Performance

Performance is measured by self-report on net profit per hour (income) from selling activities. Self-reported performance has been shown to be reliable in previous sales force research [3].

The investigated companies are devoted to different product categories. Each category may strongly identify distributor's characteristics and their communication strategies with customers. Euromonitor International identifies seven categories within direct selling industry. The statistics of these categories is shown in the Table 5. In our research, we use six product categories excluding "Home Improvement and Gardening".

Figure 3 visualizes closeness of communication tactics of companies. The graph (Fig. 3) is edge-weighted. Weight of the edge is measured by Euclidean norm which assigns to each vector the length of its arrow. Thickness of the arrows between companies shows similarity. Thicker tie means more similar tactics between pair of the companies.

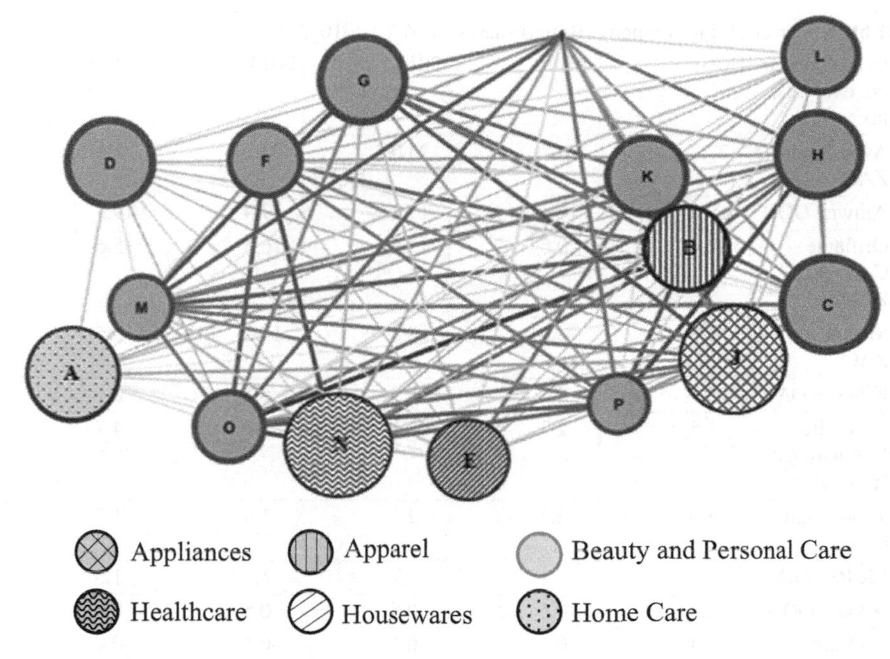

Fig. 5 Education level and distribution tactics similarity

Nodes would have different attributes. In Fig. 3, size of the nodes reflects display performance. Coloring of the node shows product category (Table 5).

As it is seen from the Fig. 3 the highest productivity has got company L from "beauty and personal care" product category. At the same time, this company has got weak tights with other companies therefore it has got unique tactic (total distance = 24.31). Group of companies from "beauty and personal care" product category with low performance from F, H, M,O have got total distance lower than 20.

Comparing communication tactics of companies from "beauty and personal care" product category (Table 3), we see that company L has got the highest frequency of using channel "Friends". Tactic of the company A from "home care" product category is also outstanding (with total distance=24.03).

We may suggest that unique tactic provides better performance. We calculated Pearson correlation coefficient between total distance and performance. For 10 companies from "beauty and personal care" industry, correlation coefficient is equal to 0.47. According to Chaddock scale the correlation is positive but weak.

Age. In Fig. 4, size of nodes reflects an average age of the distributors within the company. The bigger diameter the higher is the age. Characteristic of age has no much influence on the companies' tactic. Companies E and J with the highest age have no similarities in tactics.

Table 6 Direct selling company (Brand) shares: % Value 2010–2014

% retail value rsp excl sales tax	2010	2011	2012	2013	2014
Avon Products ZAO	27.2	24.7	23.0	23.0	21.8
Amway OOO	17.0	17.5	19.2	19.4	19.3
Oriflame Cosmetics ZAO	22.4	20.5	17.6	16.5	15.6
Mary Kay ZAO	8.8	8.6	8.8	10.0	10.5
Faberlic OAO	3.5	5.2	5.9	6.1	6.1
Herbalife International RS OOO	1.5	2.3	3.1	3.9	4.7
Tupperware OOO	3.0	2.3	2.2	1.9	1.7
LR Rus OOO	–	–	0.7	1.3	1.6
Nikken OOO	0.6	0.6	0.6	0.5	0.4
Nu Skin Russia	0.4	0.3	0.4	0.4	0.4
Others	10.9	13.6	14.6	13.3	14.5
Total	100.0	100.0	100.0	100.0	100.0

Source Euromonitor International from official statistics, trade associations, trade press, company research, trade interviews, trade sources.

Education level. In Fig. 5, size of nodes reflects an average education level of the distributors within the company. It is seen that education level is almost equal for all companies and is equivalent to higher education. Company J which is selling appliances has got distributors with the higher level of education.

Market share. Companies within research sample have got different size and market share. Table 6 presents top 10 companies of the industry. All of them are presented in our research. Totally they amount 85% of direct selling market in Russia. As it is seen from the table, there is a strong difference in market share between companies.

Figure 6 help us to analyze if company's market share has got effect on tactic. Market leaders from "beauty and personal care" companies H and P have got similarities in tactics, but are different to A company tactics from "home care" product category. But at the same time, tactics of big companies are alike to small companies. So, there is no significant difference in tactics of small and big companies.

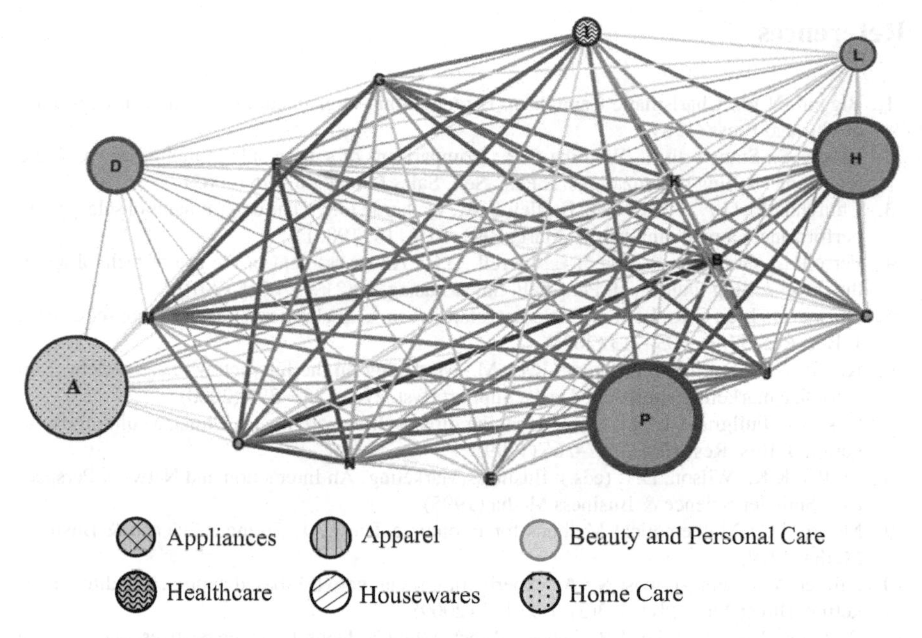

Fig. 6 Market share and distribution tactics similarity

7 Conclusions

Company success strongly depends on effective communication with customers of distributor's network. The above network representation of multidimensional statistics provides us with a deeper understanding of idiosyncrasies of distribution communication tactics between companies with different characteristics.

The main contribution of the paper is in application of network approach and Euclidean norm concept for multidimensional analysis of distribution tactics of FMCG companies. The study simultaneously estimates usage of eight communication channels applying omni-channel approach. Empirical application of the approach for analysis of distribution tactics of FMCG companies shows that it is useful for analysis of eight-dimensional tactics of numerous companies from six different product categories.

Acknowledgements The study has been funded by the Russian Academic Excellence Project '5-100'.

References

1. Biggart, N.W.: Charismatic Capitalism: Direct Selling Organizations in America. University of Chicago Press (1989)
2. Brodie, S., Stanworth, J., Wotruba, T.R.: Comparisons of salespeople in multilevel vs. single level direct selling organizations. J. Pers. Sell. Sales Manag., 67–75 (2002)
3. Churchill, Jr., G.A., Ford, N.M., Hartley, S.W., Walker Jr, O.C.: The determinants of salesperson performance: a meta-analysis. J. Mark. Res., 103–118 (1985)
4. Ferrell, L., Gonzalez-Padron, T.L., Ferrell, O.C.: An assessment of the use of technology in the direct selling industry. J. Pers. Sell. Sales Manag. **30**(2), 157–165 (2010)
5. Grewal, D., Iyer, G.R., Levy, M.: Internet retailing: enablers, limiters and market consequences. J. Bus. Res. **57**(7), 703–713 (2004)
6. Kiang, M.Y., Raghu, T.S., Shang, K.H.-M.: Marketing on the Internet–who can benefit from an online marketing approach? Decis. Support Syst. **27**(4), 383–393 (2000)
7. Luk, S.T., Fullgrabe, L., Li, S.C.: Managing direct selling activities in China: a cultural explanation. J. Bus. Res. **45**(3), 257–266 (1999)
8. Möller, K.K., Wilson, D.T. (eds.): Business Marketing: An Interaction and Network Perspective. Springer Science & Business Media (1995)
9. Moore, J.C.: Mathematical Methods for Economic Theory 1. Springer Science & Business Media (1999)
10. Palmer, A., Koenig-Lewis, N.: An experiential, social network-based approach to direct marketing. Direct Mark.: Int. J. **3**(3), 162–176 (2009)
11. Peterson, R.A., Wotruba, T.R.: What is direct selling?—Definition, perspectives, and research agenda. J. Pers. Sell. Sales Manag. **16**(4), 1–16 (1996)
12. Rigby, D.: The future of shopping harvard business review. **89**(12), 64–75 (2011)
13. Sanan, D.: From the practitioner's desk: a comment on 'What Is Direct Selling?—definition, perspectives, and research agenda'. J. Pers. Sell. Sales Manag. **17**(2 Spring), 57–59 (1997)
14. Suppes, P., Krantz, D.H.: Foundations of Measurement: Geometrical, Threshold, and Probabilistic Representations. Courier Corporation (2006)
15. Verhoef, P.C., Kannan, P.K., Inman, J. J.: From multi-channel retailing to omni-channel retailing: introduction to the special issue on multi-channel retailing. J. Retailing. **91**(2), 174–181 (2015)
16. World Federation of Direct Selling Associations annual report, 2015

Machine Learning Application to Human Brain Network Studies: A Kernel Approach

Anvar Kurmukov, Yulia Dodonova and Leonid E. Zhukov

Abstract We consider a task of predicting normal and pathological phenotypes from macroscale human brain networks. These networks (connectomes) represent aggregated neural pathways between brain regions. We point to properties of connectomes that make them different from graphs arising in other application areas of network science. We discuss how machine learning can be organized on brain networks and focus on kernel classification methods. We describe different kernels on brain networks, including those that use information about similarity in spectral distributions of brain graphs and distances between optimal partitions of connectomes. We compare performance of the reviewed kernels in tasks of classifying autism spectrum disorder versus typical development and carriers versus noncarriers of an allele associated with an increased risk of Alzheimer's disease.

Keywords Machine learning · Brain networks · Classification · Kernel SVM · Graph spectra · Clustering

1 Introduction

Recently, network representation of human brains (called connectomes) has gained increasing attention in neuroscience research. One of the challenges posed by connectomics is the classification of normal and pathological phenotypes based on brain networks [1]. Mathematically, this is a problem of classifying small undirected connected graphs with uniquely labeled nodes.

A. Kurmukov (✉) · Y. Dodonova · L.E. Zhukov
National Research University Higher School of Economics, Moscow, Russia
e-mail: kurmukovai@gmail.com

Y. Dodonova
e-mail: ya.dodonova@mail.ru

L.E. Zhukov
e-mail: lzhukov@hse.ru

V.A. Kalyagin et al. (eds.), *Models, Algorithms, and Technologies for Network Analysis*, Springer Proceedings in Mathematics & Statistics 197,
DOI 10.1007/978-3-319-56829-4_17

We start this paper with a brief overview of pitfalls common to all machine learning studies on neuroimaging data. We describe how brain networks can be constructed based on magnetic resonance images (MRI) and discuss why these networks differ from graphs arising in other application areas of network science, such as chemistry or molecular biology. We next focus on a kernel approach to classification of brain networks. We adopt kernels previously described in other contexts and also review kernels proposed in our previous studies specifically for brain networks. We compare performance of these kernels based on two real-life datasets of structural connectomes.

2 Machine Learning Application to Neuroimaging Data

Machine learning based on neuroimaging data is becoming increasingly popular; until recently, group-level statistical comparisons dominated the field. A paper [2] discusses this fundamental shift in paradigm and also highlights some pitfalls of neuroimaging-based machine learning studies. For example, these include normal anatomical inter-individual variability which can mask disease-related changes, or normal inter-individual variation in cognitive reserve which adds a lot of uncertainty to the reference standards that are based on clinical diagnoses. Also, important part of variability in neuroimaging data stems from patient selection, inter-scanner variability, and data preprocessing.

A caveat of neuroimaging-based machine learning studies is also a dysbalance between a dimensionality of the feature space and the number of subjects, and hence the problem of data reduction and a high risk of overfitting. A review [3] gives a good idea of the sample sizes typical for machine learning studies in the field of neuroscience. The authors [3] summarize 118 studies that used machine learning algorithms to predict psychiatric diagnoses based on neuroimaging data. Sample sizes in a majority of those studies did not exceed 100 participants, and most of the studies were based on less than 50 participants.

Finally, a most recent comprehensive review of neuroimaging-based single subject prediction of brain disorders can be found in [4]. Based on the analysis of more than 200 papers, the authors discuss several biases common for neuroimaging-based machine learning studies, such as a feature selection bias and an issue of hyperparameter optimization. Again, the authors [4] emphasize that the main bottleneck of this field is the limited sample size.

Importantly, the majority of studies in the area deal with voxel-level and region-level features. The former include features that are extracted at the level of individual voxels, such as voxel brightness or fractional anisotropy computed based on diffusion tensor imaging (DTI). Region-based features (e.g., region volumes or region average thicknesses) are derived by parceling brain images into zones, for example, on the basis of a standardized brain atlas.

However, there exists an alternative way of representing human brains that makes full usage of network science concepts and ideas. We discuss this approach (called connectomics) in the next section.

3 Network Representation of a Human Brain

A term connectome was proposed by [5, 6]. It stands for a network that represents brain regions and their interconnections. For very simple organisms, such as Caenorhabditis elegans, these connections can be modeled at the level of individual neurons. For human brains, connectomes represent aggregated neural pathways at the macroscopic scale. For a review of this rapidly evolving research area, we refer to [1].

To produce human structural connectomes, brain gray matter is identified on MRI scans using a segmentation algorithm and is next parceled into regions according to a brain atlas. These regions are the nodes of the constructed network. White matter streamlines are detected using a tractography algorithm. The number of streamlines that connect each pair of brain regions produces a weight for an edge between the respective nodes.

The above pipeline produces DTI-based structural connectomes. It is also possible to define so-called functional connectomes based on the fMRI scans. In this case, strength of co-activation of each pair of the regions provides weights for the edges. For a review on network modeling methods on fMRI data, we refer to [7]; we do not discuss this approach here.

Since the structural connectome is a discrete mathematical model of a human brain, the algorithms of discretization chosen in a given study largely affect the size and the structure of the resulting brain networks (e.g., see [8] for a discussion on methodological pitfalls of connectome construction). First, there is no unique way to define a set of nodes for brain graphs; we refer to [9] for a discussion on how the choice of nodal scale and gray-matter parcelation scheme affects the structure and topological properties of whole-brain structural networks.

Second, network edges can be defined differently depending on a tractography algorithm used to reconstruct white matter streamlines; for example, a paper [10] examines how outcomes of machine learning on connectomes change depending on tractography algorithms underlying edge reconstruction.

Regardless of a particular algorithm used to produce network edges, raw edge weights in the resulting structural connectomes are proportional to the number of detected streamlines. A researcher next makes a choice on whether to work with unweighted or weighted networks. The former approach implies that all raw weights are binarized. Given an undirected weighted graph with n nodes, let A be the $n \times n$ adjacency matrix with entries a_{ij}, where a_{ij} is the weight between the respective nodes. Unweighted graph is produced by:

$$a_{ij}^{binarized} = 1 if a_{ij} > 0,\ 0\, else. \quad (1)$$

Sometimes a threshold is set to a nonzero value to eliminate low weights. Alternatively, a threshold can be set different across participants, while the sparsity of the resulting networks is fixed across all brains (e.g., the authors of [11] compute graph metrics for the unweighted networks within a range of sparsity levels and next average the obtained values).

When a study analyzes weighted brain networks, normalization of connectivity matrices is recommended [12, 13]. This is because raw number of streamlines is known to vary from individual to individual and can be affected by fiber tract length, volume of cortical regions, and other factors. Normalization itself can involve geometric properties such as volumes of the cortical regions or physical path lengths between the regions (e.g., [12, 13]), or be purely based on topological properties of the networks (e.g., [14, 15]). A paper [16] examines how topological and geometric normalizations of brain networks affect the predictive quality of machine learning algorithms run on these networks. The results of [16] suggest that a combination of both topological and geometric normalizations is the most informative.

To sum up, there is certainly some ambiguity in how structural connectomes should be constructed from DTI scans. However, regardless of the particular aspects of the network reconstruction pipeline, the resulting brain graphs share some important properties. These are usually small undirected connected networks. The vertices are labeled according to brain regions, and a set of uniquely labeled vertices is the same across different connectomes constructed with the same atlas. The networks are spatially embedded: vertices are localized in 3D space, and edges have physical lengths. In what follows, we discuss how machine learning algorithms can be applied to these objects.

4 Machine Learning on Brain Networks

Hence, a problem of classifying scans of normal and pathological brains transforms into a problem of classifying the respective brain networks. Mathematically, this is a task of pattern recognition on graphs; however, it differs from a more usual understanding of machine learning on graphs. More commonly, a graph itself becomes an object defining a metric between the vertices, and machine learning algorithms are run on vertices or neighborhoods (e.g., algorithms aiming at link prediction in social networks). Connectomics poses a different challenge: small brain graphs are now examples of classes to be distinguished by an algorithm. In this section, we provide a formal problem statement and discuss how it can be tackled.

4.1 Problem Statement

Let G_i be a brain network, y_i be a class label, $y_i \in \{0, 1\}$ throughout this study. Given a training set of pairs (G_i, y_i) and the test set of input objects G_j, the task is to make a best possible prediction of the unknown class label y_j. In what follows, we use G to denote a brain graph, either unweighted or weighted, and A to denote the respective adjacency matrix which includes values from $\{0, 1\}$ if the graph is unweighted or holds edge weights if the graph is weighted. We consider the classification problem for both unweighted and weighted networks, and make special remarks on the work of the algorithms in these two cases when needed.

In some sense, this problem is similar to the problem of classifying molecules that arises in chemistry and molecular biology. A paper [17] describes some benchmark datasets from that subject area and the respective tasks, for example, a task of assigning protein molecules to a class of enzymes or nonenzymes, predicting whether or not a given molecule exerts a mutagenic effect, or whether or not a given chemical compound is cancerogenic. Each molecule or compound is modeled as a graph, with the nodes representing atoms and the edges representing bonds between the atoms; each node is labeled with its atom type.

Similarly to brain networks, molecules are small connected graphs which should be assigned a class label. The major difference between the two problems is that each node in brain networks has a unique label, and hence the problem of graph isomorphism does not arise in connectomics. This means that machine learning algorithms shown to be useful in other subject areas are to be modified to be valid for classifying brain networks; in Sect. 5, we show how this can be accommodated. Besides, an important prerequisite of classifying brain networks is that all brain graphs have the same number of nodes and the same set of node labels; in Sect. 5.5, we show how this very specific property of brain networks can be used to develop machine learning algorithms on brain graphs.

4.2 A Kernel Approach

The most obvious approach to machine learning within these settings would be to adopt some strategy of graph embedding and transform adjacency matrices into vectors from $\mathbb{R}^p$ because most classifiers work with this type of input objects. One could vectorize a matrix by taking the values of its upper triangle (the so-called "bag of edges") or compute some local or global graph metrics and use them as feature vectors. For an excellent example of a study within this framework, we refer to [15]. The authors work with the "bag of edges" and also compute edge betweenness centralities, network efficiency, clustering coefficients, and some other topological metrics and classify different sex and kinship groups based on these features.

In this paper, we focus on a different approach that defines kernels on structured data directly and hence allows for classifying brain networks without embedding brain graphs into a real vector space. This is possible due to an important property of the SVM classifier to accept any input objects, not necessarily vectors from $\mathbb{R}^p$ [18, 19]. This means that any positive semi-definite function $K(\mathbf{x}_i, \mathbf{x}_j) : \mathbb{X}^2 \to \mathbb{R}$ on the input data $\mathbb{X}$ can be used as a kernel for the SVM classifier provided that:

$$\sum_{i=1}^{n}\sum_{j=1}^{n} K(\mathbf{x}_i, \mathbf{x}_j)c_i c_j \geq 0$$

for any $(x_1, x_2, \ldots, x_n) \in \mathbb{X}$ and any coefficients $(c_1, c_2, \ldots, c_n) \in \mathbb{R}$. There are no constraints on the structure of the input data $\mathbb{X}$.

In the next sections, we review several kernels that can be useful for classifying brain networks and compare their performance based on two real-life datasets.

5 Kernels on Brain Networks

Below, we review some kernels that can be useful for a task of classifying brain networks. Recall that we only deal with structural connectomes which represent anatomical connections between brain regions. It is also possible to define functional connectomes, for which the elements of weighted adjacency matrices are the correlations between time series and the respective brain regions activation; as such, certain specific kernel methods can be developed for this particular type of input data (e.g., those that account for the geometry of the manifold of the positive definite correlation matrices [20]). In this sense, structural connectomes are more problematic as they do not lie in a specific space with known properties. In what follows, we only discuss kernel methods applicable for the analysis of structural connectomes.

We discuss two approaches to producing graph kernels. The first approach defines a kernel function that generates a positive semi-definite Gram matrix. A second approach introduces a function quantifying a distance between graphs and next obtains a kernel by exponentiating this distance.

5.1 Random Walk Kernel

We first consider a walk kernel described in [18], which computes the number of walks common for each pair of graphs. Since all brain graphs have the same set of uniquely labeled nodes Γ, we modify the walk kernel as described below.

The walk kernel is now computed on a graph G_* which is a minimum of G G'. The graph G_* has the same set of nodes $\Gamma_* = \Gamma$ and an adjacency matrix A_*:

$$A_* = a_{*ij} = \{\min(a_{ij}, a'_{ij}) : a_{ij} \in A, a'_{ij} \in A'\}. \quad (2)$$

Note that Eq. (2) produces a correct minimum graph regardless of whether the adjacency matrix A is unweighted or holds the edge weights.

We compute the walk kernel on G and G' by:

$$K_{walk}(G, G') = \sum_{i,j=1}^{|\Gamma_*|} [\sum_{k=0}^{\infty} \mu_k A_*^k]_{ij}. \tag{3}$$

We set $\mu_k = \mu^k$. Hence, the (3) becomes:

$$K_{walk}(G, G') = \sum_{i,j=1}^{|\Gamma_*|} [\sum_{k=0}^{\infty} \mu^k A_*^k]_{ij} = \sum_{i,j=1}^{|V_*|} [(I - \mu A_*)^{-1}]_{ij} \tag{4}$$

To ensure convergence, μ must be lower than the inverse maximal eigenvalue. In this paper, we report results for μ set to 0.95 times the inverse maximal eigenvalue of A_*; lower values of μ tried in preliminary studies result in slightly worse classification quality. Conceptually, the factor μ downweights longer walks and makes short walks dominate the graph similarity score. A paper [17] discusses this effect.

In addition to sensitivity to the length of walks taken into account, walk kernel suffers from the so-called tottering effect [21]. Since walks allow for repetitions of nodes and edges, the same fragment is counted repeatedly in a graph similarity measure. In undirected graphs, a random walk may start tottering on a cycle or even between the same two nodes in the product graph, leading to an artificially high graph similarity score even when the structural similarity between two graphs is minor.

5.2 *Kernel on Shortest Path Lengths*

Second, we consider a kernel on shortest path lengths. Kernels on shortest path lengths are proposed in [17] as an alternative to random walk kernel that overcomes its shortcomings discussed above. The authors [17] define a kernel on graphs that compares paths instead of walks. In this study, we only use one version of a kernel based on paths, with some preliminary modification aiming to account for unique node labels of brain networks.

For a graph G with a set of uniquely labeled nodes Γ, a matrix of shortest path lengths is given by:

$$\Upsilon_{ij} = \upsilon(\gamma_i, \gamma_j), \tag{5}$$

where γ_i and γ_j are the nodes of the graph G and $\upsilon(\gamma_i, \gamma_j)$ is the length of the shortest path between these nodes (weighted or unweighted, depending on the nature of graph G).

We next define a path kernel by:

$$K_{path}(\Upsilon, \Upsilon') = \sum_{\substack{\upsilon_{ij} \in \Upsilon \\ \upsilon'_{ij} \in \Upsilon'}} K_1(\upsilon_{ij}, \upsilon'_{ij}), \tag{6}$$

where $K_1(v_{ij}, v'_{ij})$ is a kernel on pairs of paths from G and G'. For the later, we use a polynomial kernel $K_1(\mathbf{x}, \mathbf{x}')$ given by:

$$K_{poly}(\mathbf{x}, \mathbf{x}') = (\langle \mathbf{x}, \mathbf{x}' \rangle + c)^p, \quad p = 2. \tag{7}$$

For a general definition of path kernels and proof of their positive definiteness, we refer to [17].

5.3 Distance-Based Kernels: L_1 and L_2 Norms

The above methods produce Gram matrices on graphs straightforwardly. An alternative approach is to introduce a distance between graphs and produce a kernel based on this distance measure.

Let G and G' be the networks and $\omega(G, G')$ be a distance between these networks. We build a graph kernel K using the distance ω as follows:

$$K(G, G') = e^{-\alpha\omega(G,G')} \tag{8}$$

Positive semi-definiteness of this kernel is guaranteed when ω is a metric. A paper [22] discusses kernels which are not necessarily positive semi-definite, namely those for which triangle inequality does not hold for a distance measure ω in (8). The authors claim that these kernels can always be made positive definite by an appropriate choice of the parameter α; however, forcing a kernel to be positive definite reduces its expressiveness and diminishes classification accuracy. In this study, we vary the parameter α for all distance-based kernels, including those using true metric ω.

We first define distances between networks via the L_1 and L_2 norms between the respective adjacency matrices. For two networks G and G' with $n \times n$ adjacency matrices $A = \{a_{ij}\}$ and $A' = \{a'_{ij}\}$ (either unweighted or weighted), an L_1 distance is given by:

$$\omega_{L_1}(G, G') = \sum_{i=1}^{n}\sum_{j=1}^{n} |a_{ij} - a'_{ij}| \tag{9}$$

An L_2 (Frobenius) norm is defined by:

$$\omega_{L_2}(G, G') = \sqrt{\sum_{i=1}^{n}\sum_{j=1}^{n} (a_{ij} - a'_{ij})^2} \tag{10}$$

We next produce kernels (8) based on these distance measures. This is the simplest possible way to define a distance between the adjacency matrices. In the next sections, we discuss more sophisticated procedures aiming to quantify pairwise distances between networks.

5.4 Kernels on Distances Between Spectral Distributions

Studies [23, 24] proposed to measure similarity between brain networks based on distances between spectral distributions of the respective graphs. An idea behind spectral-based kernels is that graph eigenvalue distributions capture important information about network structure and hence might be useful for a task of classifying networks.

To construct spectral-based kernels, we use spectra of the normalized graph Laplacians. Let D be a diagonal matrix of weighted node degrees:

$$d_i = \sum_j a_{ij}. \tag{11}$$

The graph Laplacian matrix is given by:

$$L = D - A, \tag{12}$$

The normalized graph Laplacian is given by:

$$\mathcal{L} = D^{-1/2} L D^{-1/2} \tag{13}$$

Normalized Laplacians are correctly defined by (13) regardless of whether the graphs are unweighted or weighted, provided that for weighted graph the matrix A holds edge weights. The eigenvalues of the normalized Laplacians are always in range from 0 to 2. We refer to [25] for theory on the normalized Laplacian spectra and to [26] for examples of the eigenvalue distributions of the normalized Laplacians in structural brain networks of the cat, macaque, and Caenorhabditis elegans.

A paper [23] defines distances between brain networks via the information-theory based measures of difference between spectral distributions. A motivation behind this approach is that we are most interested in comparing shapes of the distributions of eigenvalues rather than the vectors of eigenvalues per se. This is because multiplicity of particular eigenvalues and specific peaks in their distributions capture important information about graph structure [25].

To quantify distance between distributions, we use the Kullback–Leibler (KL) divergence and the Jensen–Shannon (JS) divergence. For two probability distributions with densities $p(x)$ and $q(x)$ the KL divergence is:

$$KL(p||q) = \int_{-\infty}^{\infty} p(x) log \frac{p(x)}{q(x)} dx \tag{14}$$

The Kullback–Leibler kernel [27] is obtained by exponentiating, the symmetric KL divergence:

$$K_{KL}(p, q) = e^{-\alpha(KL(p||q)+KL(q||p))} \tag{15}$$

The JS divergence [28] is:

$$JS(p||q) = \frac{1}{2}(KL(p||r) + KL(q||r)), \tag{16}$$

where $r(x) = \frac{1}{2}(p(x) + q(x))$.

We compute Jensen–Shannon kernel by:

$$K_{JS}(p, q) = e^{-\alpha\sqrt{JS(p||q)}} \tag{17}$$

The KL and JS kernels work with the probability density functions restored from the samples. In [23], we split the entire range of eigenvalues into equal intervals (bins) and computed frequencies within each bin as a proxy for the underlying probabilities. However, the results of the entire classification pipeline were highly sensitive to the choice of the number of bins used to reconstruct density. In this study, we overcome this shortcoming by applying kernel density reconstruction prior to computation the KL and JS divergences. We use the Gaussian kernel and produce the values:

$$f(x) = \sum_{s_j} \frac{1}{\sqrt{2\pi\sigma^2}} exp(-\frac{|x - s_j|^2}{2\sigma^2}), \tag{18}$$

where s_j is the j-th eigenvalue of $\mathcal{L}$. To compute this, we use the statmodels [29] function for univariate kernel density estimation, which is a fast Fourier transform-based implementation that has an advantage of automatic selection of optimal bandwidth according to the Silverman's rule. We next compute the kernels (15) and (17) based on these values.

There also exists a different approach that compares spectral distributions directly based on the vectors of eigenvalues [24]. For this purpose, it uses an earth mover's distance (EMD) [30], which measures the minimum cost of transforming one sample distribution into another. Provided that each distribution is represented by some amount of dirt, EMD is the minimum cost of moving the dirt of one distribution to produce the other. The cost is the amount of dirt moved times the distance by which it is moved.

Let $\{s_1^i, \ldots, s_n^i\}$ be the eigenvalues of the normalized Laplacian spectrum $\mathcal{S}_i$. We put an equal measure 1/n to each point s_k^i on a real line. Let f_{kl} be the flow of mass between the points s_k^i and s_l^j. The EMD is the normalized flow of mass between sets $\mathcal{S}_i = \{s_1^i, \ldots, s_n^i\}$ and $\mathcal{S}_j = \{s_1^j, \ldots, s_n^j\}$ that minimizes the overall cost:

$$emd(\mathcal{S}_i, \mathcal{S}_j) = \underset{F=\{f_{kl}\}}{\operatorname{argmin}} \frac{\sum_{k,l} f_{kl}|s_k^i - s_l^j|}{\sum_{k,l} f_{kl}}, \tag{19}$$

with the constraints: $f_{kl} \geq 0, \quad \sum_{k=1}^n f_{kl} = 1/n, \quad \sum_{l=1}^n f_{kl} = 1/n.$

A EMD-based kernel is next computed by (8) using (19) as a measure of distance between the respective graphs.

5.5 Kernels on Distances Between Network Partitions

The last group of graph kernels analyzed in this study quantifies similarity between brain networks based on whether or not their nodes cluster into similar communities [31]. Partition-based kernels make the full use of the uniqueness of node labels in brain networks and the identity of label sets across different brains. These kernels are based on the idea that brain networks belonging to a same class produce partitions that are more similar than those obtained for networks from different classes.

Similarly to [31], this study uses three algorithms to obtain partition of each brain network: Newman leading eigenvector method [32], Louvian method [33], and Greedy modularity optimization [34]. All these methods use modularity as a function to be optimized. *Modularity* [34] is a property of a network and a particular division of that network into communities. It measures how good is the division in the sense that whether there are many edges within communities and only a few between them. Modularity Q is given by:

$$Q = \frac{1}{2m} \sum_{ij} [a_{ij} - \frac{d_i d_j}{2m}] \delta(i, j), \quad (20)$$

where a_{ij} is an element of a graph adjacency matrix, m is a total number of edges in a given graph, d_i, d_j—degrees of nodes i and j as defined by (11).

Louvain algorithm is a two-step iterative procedure. It starts with all nodes put in separate clusters. Next, for each node i and its neighbors j, the algorithm computes gain in modularity that would take place after removing i from its cluster and placing it to a cluster of j; after repeating for all neighbors j, i is placed in the cluster where gain in modularity is maximal. This process repeats until there is no such node i for which its movement to another cluster produces gain in modularity. The second step of the algorithm builds a new weighted graph wherein nodes are final clusters from the previous step and an edge between two nodes represents the sum of edges between two corresponding clusters at the previous step. Once the second step is over, the algorithm reapplies the first step and iterates.

The Newman leading eigenvector method uses normalized graph Laplacian given by (12). It starts with all nodes placed in a single cluster; different nodes next get their labels according to a sign of the respective values of a Laplacian eigenvector corresponding to the second smallest eigenvalue. The procedure repeats for each cluster till convergence. Greedy modularity optimization method is another division clustering approach which allows for fast detecting communities in large graphs or sets of many small graphs.

All these partition algorithms are defined for both unweighted and weighted graphs. We next estimate pairwise similarity of partitions of different brain networks using the adjusted Rand score (ARI). Let $U = \{U_1, U_2, \dots U_l\}$ and $V = \{V_1, V_2, \dots V_k\}$ be partitions of two networks G_U and G_V with the same sets of node labels, l and k be the number of clusters in the partitions U and V, respectively. To define ARI between these partitions, we construct a contingency table:

U, V	V_1	V_2	$\dots$	V_k	sum
U_1	ν_{11}	ν_{12}	$\dots$	ν_{1k}	a_1
U_2	ν_{21}	ν_{22}	$\dots$	ν_{2k}	a_2
$\vdots$	$\vdots$	$\vdots$	$\ddots$	$\vdots$	$\vdots$
U_l	ν_{l1}	ν_{l2}	$\dots$	ν_{lk}	a_l
sum	b_1	b_2	$\dots$	b_k	

Here ν_{ij} denotes a number of objects common between U_i and V_j. ARI is then given by:

$$ARI(U, V) = \frac{\sum_{i,j} \binom{\nu_{ij}}{2} - \left[\sum_i \binom{a_i}{2} \sum_j \binom{b_i}{2}\right] / \binom{\nu}{2}}{\frac{1}{2}\left[\sum_i \binom{a_i}{2} + \sum_j \binom{b_i}{2}\right] - \left[\sum_i \binom{a_i}{2} \sum_j \binom{b_i}{2}\right] / \binom{\nu}{2}}. \tag{21}$$

ARI takes the value of 1 when the partitions are identical and values close to 0 in case of random labeling. We thus define a distance $\omega(G_U, G_V)$ between networks G_U and G_V by:

$$\omega(G_U, G_V) = 1 - ARI(U, V), \tag{22}$$

Hence, networks with the same partitions have zero distance, and the maximum distance is close to 1. We next produce three kernels (8) based on these pairwise distances, one for each algorithm of clustering brain networks.

6 Summary: Methods

We compare performance of the kernels described in the previous section based on two tasks of classifying brain networks. In this section, we overview the classification pipeline and describe a metric used to compare performance of different kernels; in the next section, we describe the tasks and the datasets.

6.1 Classification Pipeline

Figure 1 summarizes our classification pipeline. We deal with brain networks that represent different phenotypes (i.e., normal and pathological brains). We compute Gram matrices between these brain networks using the following kernels:

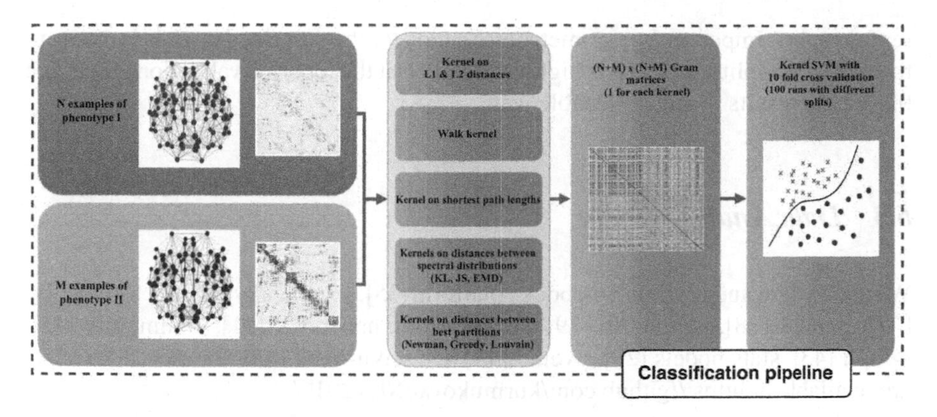

Fig. 1 Classification pipeline

- Random walk (RW) kernel [18]
- Shortest path length (SPL) kernel [17]
- L_1-distance kernel
- L_2-distance kernel
- Kullback–Leibler (KL) kernel [27]
- Jensen–Shannon (JS) kernel [23]
- Earth mover's distance (EMD) kernel [24]
- Newman partition (NP) kernel [31]
- Louvain partition (LP) kernel [31]
- Greedy partition (GP) kernel [31]

We next feed these Gram matrices to an SVM classifier, train it on part of a sample and make prediction for an unseen part of a sample. In computation of distance-based kernels, we vary the values of α in the range from 0.01 to 10. The penalty parameter of the SVM classifier varies from 0.1 to 50. We report the results for models with the optimal values of α and the penalty parameter.

6.2 Classification Quality Evaluation

We use the area under the receiver operating characteristic curve (ROC AUC) to assess the predictive quality of models with different kernels. We run all models with tenfold cross-validation and combine predictions on test folds to evaluate the quality of prediction on the entire sample. For each model, we repeat this procedure 100 times with different tenfold splits, thus producing 100 ROC AUC values.

Note that datasets under consideration are too small to divide them into three parts (train, validation and test) for parameter estimation. To deal with this, we find optimal values of the parameters based on 10 different tenfold splits (with random states fixed for splitting) and evaluate models with optimal parameter values based on 100 other tenfold splits. Importantly, parameter estimation for the described models is robust to the particular splitting, and it is highly unlikely that the reported validation procedure biased the results in any noticeable way.

6.3 Data Analysis Tools

We use Python and IPython notebooks platform [35], specifically NumPy [36], SciPy [37], pandas [38], matplotlib [39], seaborn [40], networkX [41], community [42], igraph [43], statsmodels [29], pyemd [44] and scikit-learn [45] libraries. All scripts are available at https://github.com/kurmukovai/NET2016/.

7 Data

We compare the performance of the kernels described in Sect. 5 based on two datasets of precomputed matrices of structural connectomes. We describe the datasets in this section and also provide some relevant information on the resulting networks.

7.1 Datasets

UCLA Autism dataset (UCLA Multimodal Connectivity Database [11, 46]) includes DTI-based connectivity matrices of 51 high-functioning autism spectrum disorder (ASD) subjects (6 females) and 43 typically developing (TD) subjects (7 females). Average age (age standard deviation) is 13.0 (2.8) for ASD group and 13.1 (2.4) for TD group. Nodes of brain networks are defined using a parcelation scheme by Power et al. [47] which is based on a meta-analysis of fMRI studies combined with whole-brain functional connectivity mapping. This approach produces 264 equal-size brain regions and thus 264×264 connectivity matrices. Network edges are produced based on deterministic tractography performed using the fiber assignment by continuous tracking (FACT) algorithm [48]; edge weights are proportional to the number of streamlines detected by FACT.

UCLA APOE-4 dataset (UCLA Multimodal Connectivity Database [46, 49]) includes DTI-based connectivity matrices of carriers and noncarriers of the APOE-4 allele associated with the higher risk of Alzheimer's disease. The sample includes 30 APOE-4 noncarriers, mean age (age standard deviation) is 63.8 (8.3), and 25 APOE-4 carriers, mean age (age standard deviation) is 60.8 (9.7). Each brain is partitioned into 110 regions using the Harvard-Oxford subcortical and cortical probabilistic atlases as implemented in FSL [50]. Therefore, this dataset includes 110×110 connectivity matrices. Network edges are obtained using the FACT algorithm [48]. Raw fiber counts in these matrices are adjusted for the unequal region volumes by scaling each edge by the mean volume of its two adjacent regions.

The authors of both datasets only report the results of statistical group comparison based on graph metrics. Hence, there is no publicly available machine learning baselines for these datasets.

7.2 Edge Weights

For each classification task, we evaluate performance of all kernels on both unweighted and weighted brain networks. We produce unweighted brain networks by (1). In this case, each network contains information only on presence or absence of edges between nodes, and all edges carry equal weights.

To produce weighted brain networks, we take the original edge weights that represent streamline count between each pair of brain regions and scale them by the physical distances between the respective regions:

$$a_{ij}^{scaled} = \frac{a_{ij}}{\lambda_{ij}}, \tag{23}$$

where a_{ij} is the original weight of the edge between the nodes i and j, and λ_{ij} is the Euclidean distance between centers of the regions i and j. The distances are computed based on the standard Montreal Neurological Institute (MNI) coordinates of region centers.

To enhance between-subject comparison, we next normalize the obtained weights by:

$$a_{ij}^{normed} = \frac{a_{ij}^{scaled}}{\sum_{i,j} a_{ij}^{scaled}}. \tag{24}$$

Note that this latter scaling does not affect the kernels that are based on normalized Laplacian spectra and the partition-based kernels.

We report classification results for both weighted and unweighted connectivity matrices.

8 Results: Kernel Comparison

Figure 2 compares performance of the SVM classifier with different kernels in a task of classification typical development versus autism spectrum disorder. Figure 3 provides results for a task of classifying of carriers versus noncarriers of an allele associated with an increased risk of Alzheimer's disease.

First, the results show that the classifiers run on weighted brain networks clearly outperform those run on unweighted brain graphs (the only exception is an SVM

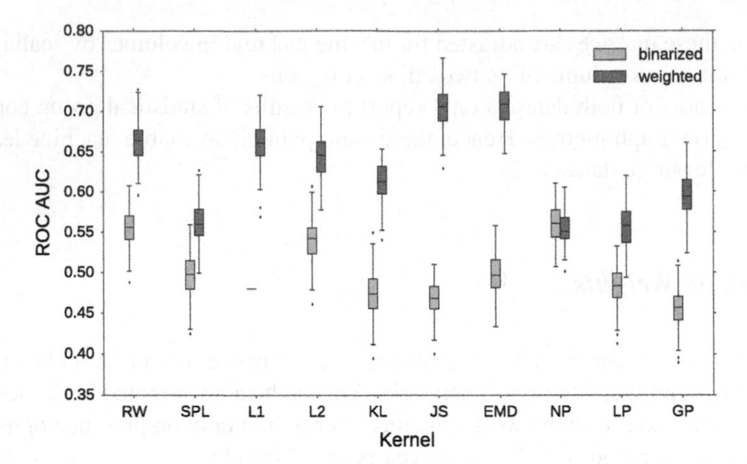

Fig. 2 Classification of typical development versus autism spectrum disorder; boxplots show ROC AUC values over 100 runs of the algorithm with different splits into train and test samples; abbreviations of the kernels are the same as in Sect. 6.1: RW—random walk, SPL—shortest path length, L1—L_1-distance, L2—L_2-distance, KL—Kullback–Leibler, JS—Jensen–Shannon, EMD—earth mover's distance, NP—Newman-based partition, LP—Louvain-based partition, GP—greedy partition

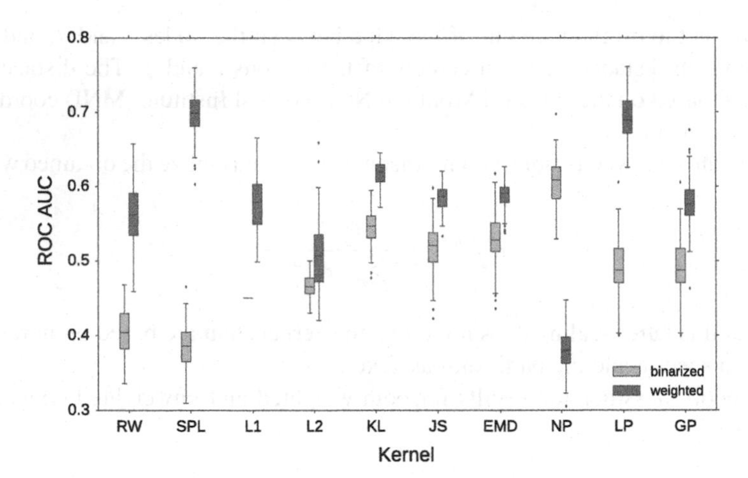

Fig. 3 Classification of carriers versus noncarriers of an allele associated with an increased risk of Alzheimer's disease; boxplots show ROC AUC values over 100 runs of the algorithm with different splits into train and test samples; abbreviations of the kernels are the same as in Sect. 6.1: RW—random walk, SPL—shortest path length, L1—L_1-distance, L2—L_2-distance, KL—Kullback–Leibler, JS—Jensen–Shannon, EMD—earth mover's distance, NP—Newman-based partition, LP—Louvain-based partition, GP—greedy partition

with the Newman-based partition kernel run on the UCLA APOE-4 dataset). This means that edge weights in human macroscale brain networks capture information important for classifying normal and pathological phenotypes. This is true regardless of whether we construct a kernel based on similarity of single edges or shortest paths, or random walks common between brain graphs, or distances between graph spectral distributions or partitions of brain networks. Importantly, edge weights in this study incorporate information on both strengths of connections (the number of streamlines detected by tractography algorithms) and their lengths (approximated by Euclidean distances between the centers of brain regions).

Second, there is no kernel (or no family of kernels) that provides the best classification quality on both datasets. Random walk kernel and L_1 and L_2 distance-based kernels do not perform satisfactorily in both classification tasks.

In a task of classifying autism spectrum disorder versus typical development, kernels based on distances between spectral distributions perform the best. There is virtually no difference in behaviors of the two kernels of this type, computed with Jensen–Shannon divergence and earth mover's distance. Spectral distributions of brain networks seem to capture some information important for distinguishing this type of pathology from typical development.

For classification of carriers versus noncarriers of an APOE-4 allele, the most expressive kernels are using comparison of shortest path lengths and the distances between Louvain-based partitions of brain networks. Interestingly, the three partition-based kernels differed in their performance, which means that the analyzed partition algorithms capture different aspects of brain network structures and thus produce distances between brain networks in a different manner.

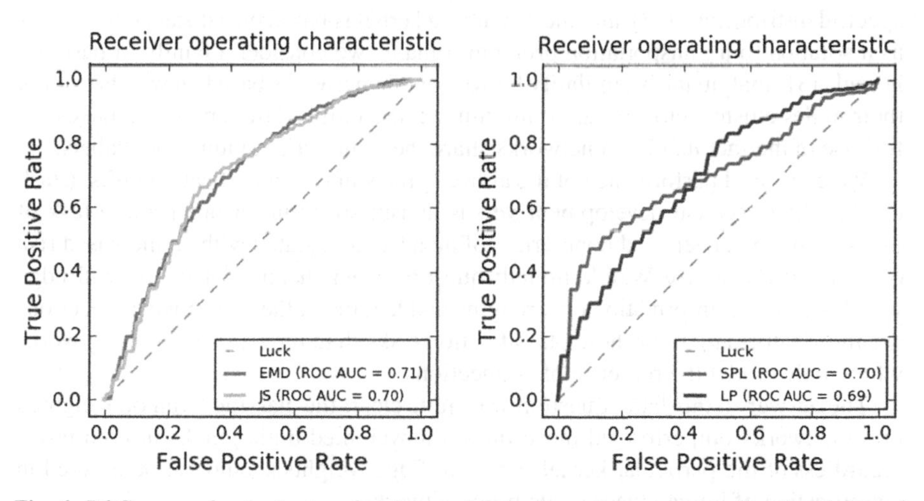

Fig. 4 ROC-curves for the best-performing models in the autism spectrum disorder (*left*) and APOE-4 allele carriers (*right*) classification tasks. Abbreviations of the kernels are the same as in Sect. 6.1: SPL—shortest path length, JS—Jensen–Shannon, LP—Louvain-based partition

For the two best models on each dataset, we plot the ROC-curves in Fig. 4. The curves are averaged over 100 repetitions of the algorithms. Interestingly, the ROC-curves do not coincide. This means that although the best-working models are close in terms of classification quality, they capture different aspects of the data and differ in terms of prediction outcome.

9 Conclusions

In this paper, we considered machine learning on macroscale human brain networks. These networks (called connectomes) represent connections between brain regions reconstructed from neuroimaging data. A question is whether connectomes can be useful in discriminating between normal and pathological brain structures, which can be considered a task of classification on graphs. We point to properties of brain networks that make a task of classifying connectomes differ from a task of classifying graph objects from other subject areas.

We next focus on a kernel classification approach and discuss several kernels that can be useful for machine learning on connectomes. We show how a random walk kernel [18] and a kernel based on shortest path lengths [17] can be modified to account for the uniqueness of node labels in brain graphs. We consider an approach that produces kernels based on distances between the adjacency matrices of the respective graphs and use L_1 and L_2 as the simplest examples of such distances. We next describe a family of kernels that are based on graph spectral distributions; of these, two kernels use measures that quantify information divergence between spectral distributions [23], and the remaining kernel is based on a distance that arises as a solution to a transportation problem. Finally, we consider a family of partition kernels [31] that quantify similarity between brain networks based on whether or not their nodes cluster into similar communities; hence, this latter approach makes the full use of the fact that brain networks share the same set of unique node labels.

We compared performance of the above kernels in two classification tasks: a task of classifying typical development versus autism spectrum disorder and a task of distinguishing carriers and noncarriers of an allele associated with an increased risk of Alzheimer's disease. We additionally questioned whether brain networks with edge weights carrying information on strengths and lengths of the respective connections are more informative for these classification tasks than unweighted brain networks which only model the presence of connections.

The answer to this latter question was quite clear: the classifiers run on weighted brain networks outperformed those run on unweighted brain graphs in both tasks, regardless of the particular kernel function. Edge weights should not be ignored in classification of human macroscale brain networks.

The best-performing kernels were task-specific. In a task of classifying autism spectrum disorder versus typical development, spectral distributions of brain networks seem to carry information useful for distinguishing between these two classes; the two best-performing models quantified distances between networks based on similarity in their spectral distributions. However, these kernels did not perform well in classification of carriers and noncarriers of an allele associated with an increased risk of Alzheimer's disease. In this latter task, the kernels based on shortest path lengths and the similarity in partitions of brain networks were the most expressive.

The kernels analyzed in this study seem to capture different aspects of network structures specific for normal and pathological brains. Future studies may aim at aggregating information stemming from different kernel models in order to improve the quality of machine learning on brain networks.

Acknowledgements The study was supported within the framework of the Academic Fund Program at the National Research University Higher School of Economics (HSE) in 2016 (grant #16-05-0050) and by the Russian Academic Excellence Project "5–100".

References

1. Craddock, R.C., Jbabdi, S., Yan, C.G., Vogelstein, J.T.: Imaging human connectomes at the macroscale. Nat. Methods **10**(6), 524–539 (2013)
2. Haller, S., Lovblad, K.-O., Giannakopoulos, P., Van De Ville, D.: Multivariate pattern recognition for diagnosis and prognosis in clinical neuroimaging: state of the art, current challenges and future trends. Brain Topogr. **27**(3), 329–337 (2014)
3. Wolfers, T., Buitelaar, J.K., Beckmann, C.F., Franke, B., Marquand, A.F.: From estimating activation locality to predicting disorder: a review of pattern recognition for neuroimaging-based psychiatric diagnostics. Neurosci. Biobehav. Rev. **57**, 328–349 (2015)
4. Arbabshirani, M.R., Plis, S., Sui, J., Calhoun, V.D. (2016) Single subject prediction of brain disorders in neuroimaging: Promises and pitfalls. NeuroImage (in press)
5. Hagmann, P.: From diffusion MRI to brain connectomics (Thesis). EPFL, Lausanne (2005)
6. Sporns, O., Tononi, G., Ktter, R.: The human connectome: a structural description of the human brain. PLoS Computat. Biol. **1**(4), e42 (2005)
7. Smith, S.M., Miller, K.L., Salimi-Khorshidi, G., Webster, M., et al.: Network modelling methods for FMRI. NeuroImage **54**(2), 875–891 (2011)
8. Fornito, A., Zalesky, A., Breakspear, M.: Graph analysis of the human connectome: promise, progress, and pitfalls. Neuroimage **15**(80), 426–444 (2013)
9. Zalesky, A., Fornito, A., Harding, I.H., Cocchi, L., Ycel, M., Pantelis, C., Bullmore, E.T.: Neuroimage **50**(3), 970–983 (2010)
10. Zhan, L., Zhou, J., Wang, Y., et al.: Comparison of nine tractography algorithms for detecting abnormal structural brain networks in Alzheimer's disease. Front. Aging Neurosci. **14**(7), 48 (2015)
11. Rudie, J.D., Brown, J.A., Beck-Pancer, D., Hernandez, L.M., Dennis, E.L., Thompson, P.M., et al.: Altered functional and structural brain network organization in autism. Neuroimage Clin. **2**, 79–94 (2013)
12. Bassett, D.S., Brown, J.A., Deshpande, V., Carlson, J.M., Grafton, S.: Conserved and variable architecture of human white matter connectivity. Neuroimage **54**(2), 1262–1279 (2011)
13. Hagmann, P., Kurant, M., Gigandet, X., Thiran, P., Wedeen, V.J., Meuli, R., Thiran, J.-T.: Mapping human whole-brain structural networks with diffusion MRI. PLoS One **2**(7), e597 (2007)
14. Gong, G., Rosa-Neto, P., Carbonell, F., Chen, Z.J., He, Y., Evans, A.C.: Age- and gender-related differences in the cortical anatomical network. J. Neurosci. **29**(50), 15684–15693 (2009)

15. Duarte-Carvajalino, J.M., Jahanshad, N., Lenglet, C., McMahon, K.L., de Zu-bicaray, G.I., Martin, N.G., Wright, M.J., Thompson, P.M., Sapiro, G.: Hierarchical topological network analysis of anatomical human brain connectivity and differences related to sex and kinship. Neuroimage **59**(4), 3784–3804 (2012)
16. Petrov, D., Dodonova, Y., Zhukov, L.E., Belyaev, M.: Boosting connectome classification via combination of geometric and topological normalizations. In: IEEE 6th International Workshop on Pattern Recognition in Neuroimaging (PRNI) (2016). http://dx.doi.org/10.1109/PRNI.2016.7552353
17. Borgwardt, K.M.: Graph kernels. Dissertation (2007)
18. Gartner, T.: A survey of kernels for structured data. SIGKDD Explor. **5**(1), 49–58 (2003)
19. Kashima, H., Tsuda, K., Inokuchi, A.: Marginalized kernels between labeled graphs. In: Proceedings of the International Conference on Machine Learning, pp. 321–328
20. Dodero, L., Minh, H.Q., Biagio, M.S., Murino, V., Sona, D.: Kernel-based classification for brain connectivity graphs on the Riemannian manifold of positive definite matrices. In: Proceedings of the International Symposium on Biomedical Imaging, pp. 42–45
21. Mahé, P., Ueda, N., Akutsu, T., Perret, J.-L., Vert, J.-P.: Extensions of marginalized graph kernels. In: Proceedings of the Twenty-First International Conference on Machine Learning, pp. 552–559 (2004)
22. Chan, A.B., Vasconcelos, N., Moreno, P.J. (2004) A family of probabilistic kernels based on information divergence. University of California, San Diego. Technical Report, SVCL-TR-2004-1
23. Dodonova, Y., Korolev, S., Tkachev, A., Petrov, D., Zhukov, L.E., Belyaev, M.: Classification of structural brain networks based on information divergence of graph spectra. In: 2016 IEEE 26th International Workshop on Machine Learning for Signal Processing (MLSP) (2016). http://dx.doi.org/10.1109/MLSP.2016.7738852
24. Dodonova, Y., Belyaev, M., Tkachev, A., Petrov, D., Zhukov, L.E.: Kernel classification of connectomes based on earth mover's distance between graph spectra. In: 2016 1st Workshop on Brain Analysis using COnnectivity Networks (BACON MICCAI) (2016). https://arxiv.org/abs/1611.08812
25. Chung, F.: Spectral Graph Theory (1997)
26. de Lange, S.C., de Reus, M.A., van den Heuvel, M.P.: The Laplacian spectrum of neural networks. Front. Comput. Neurosci., 1–12 (2014)
27. Moreno, P.J., Ho, P.,Vasconcelos, N.: A Kullback-Leibler divergence based kernel for SVM classification in multimedia applications. Adv. Neural Inf. Process. Syst. (2003)
28. Lin, J.: Divergence measures based on Shannon entropy. IEEE Trans. Inf. Theory **37**(14), 145–151 (1991)
29. Seabold, S., Perktold, J.: Statsmodels: econometric and statistical modeling with python. In: Proceedings of the 9th Python in Science Conference (2010)
30. Rubner, Y., Tomasi, C., Guibas, L.J.: The earth movers distance as a metric for image retrieval. Int. J. Comput. Vis. **40** (2000)
31. Kurmukov, A., Dodonova, Y., Zhukov, L.: Classification of normal and pathological brain networks based on similarity in graph partitions. In: The Sixth IEEE ICDM Workshop on Data Mining in Networks. IEEE Computer Society (to appear)
32. Newman, M.E.J.: Finding community structure in networks using the eigenvectors of matrices. Phys. Rev. E **74**, 036104 (2006)
33. Blondel, V.D., Guillaume, J.-L., Lambiotte, R., Lefebvre, R.: Fast unfolding of communities in large networks. J. Stat. Mech. Theory Exp. **10**, P10008 (2008)
34. Clauset, A., Newman, M.E.J., Moore, C.: Finding community structure in very large networks. Phys. Rev. E **70**, 066111 (2004)
35. Pérez, F., Granger, B.E.: IPython: a system for interactive scientific computing. Comput. Sci. Eng. **9**, 21–29 (2007)
36. van der Walt, S., Colbert, S.C., Varoquaux, G.: The NumPy array: a structure for efficient numerical computation. Comput. Sci. Eng. **13**, 22–30 (2011)

37. Jones, E., Oliphant, E., Peterson, P., et al.: SciPy: Open Source Scientific Tools for Python (2001). http://www.scipy.org/. Accessed 03 Jun 2016
38. McKinney, W.: Data structures for statistical computing in python. In: Proceedings of the 9th Python in Science Conference, pp. 51–56 (2010)
39. Hunter, J.D.: Matplotlib: a 2D graphics environment. Comput. Sci. Eng. **9**(3), 90–95 (2007)
40. Seaborn: v0.5.0. doi:10.5281/zenodo.12710
41. Hagberg, A.A., Schult, D.A., Swart, P.J.: Exploring network structure, dynamics, and function using NetworkX. In: Proceedings of the 7th Python in Science Conference, pp. 11–15 (2008)
42. http://perso.crans.org/aynaud/communities/api.html
43. Csardi, G., Nepusz, T.: The igraph software package for complex network research. InterJ. Complex Syst. 1695 (2006). http://igraph.org/python/
44. https://github.com/garydoranjr/pyemd
45. Pedregosa, F., Varoquaux, G., Gramfort, A., Michel, V., Thirion, B., Grisel, O., Blondel, M., Prettenhofer, P., Weiss, R., Dubourg, V., Vanderplas, J., Passos, A., Cournapeau, D., Brucher, M., Perrot, M., Duchesnay, É.: Scikit-learn: machine learning in python. J. Mach. Learn. Res. **12**, 2825–2830 (2011)
46. Brown, J.A., Rudie, J.D., Bandrowski, A., Van Horn, J.D., Bookheimer, S.Y.: The UCLA multimodal connectivity database: a web-based platform for brain connectivity matrix sharing and analysis. Front. Neuroinf. **6**, 28 (2012)
47. Power, J.D., Cohen, A.L., Nelson, S.M., Wig, G.S., Barnes, K.A., Church, J.A., Vogel, A.C., Laumann, T.O., Miezin, F.M., Schlaggar, B.L., Petersen, S.E.: Functional net-work organization of the human brain. Neuron **72**, 665–678 (2011)
48. Mori, S., Crain, B.J., Chacko, V.P., Van Zijl, P.C.: Three-dimensional tracking of axonal projections in the brain by magnetic resonance imaging. Ann. Neurol. **45**, 265–269 (1999)
49. Brown, J.A., Terashima, K.H., Burggren, A.C., et al.: Brain network local interconnectivity loss in aging APOE-4 allele carriers. PNAS **108**(51), 20760–20765 (2011)
50. Jenkinson, M., Beckmann, C.F., Behrens, T.E., Woolrich, M.W., Smith, S.M.: FSL. NeuroImage **62**, 782–790 (2012)

Co-author Recommender System

Ilya Makarov, Oleg Bulanov and Leonid E. Zhukov

Abstract Modern bibliographic databases contain significant amount of information on publication activities of research communities. Researchers regularly encounter challenging task of selecting a co-author for joint research publication or searching for authors, whose papers are worth reading. We propose a new recommender system for finding possible collaborator with respect to research interests. The recommendation problem is formulated as a link prediction within the co-authorship network. The network is derived from the bibliographic database and enriched by the information on research papers obtained from Scopus and other publication ranking systems.

1 Introduction

Nowadays, researchers have to deal with hundreds of papers to become familiar with their fields of study. However, the number of the papers exceeds human abilities to read them all. The most common way to select relevant articles is by sorting a list of all articles according to a citation index and choosing some articles from the top of the list. However, such a method does not take into account the author professional specialization. Another way of selecting suitable articles is to choose articles of well-known authors. A more advanced methods was proposed by Newman in [1, 2], where he ranked authors according to the collaboration weight or values of centrality metrics such as degree and betweenness in the co-authorship network. In [3] authors decided to cluster authors at a co-authorship network who studied a particular disease,

I. Makarov (✉) · O. Bulanov · L.E. Zhukov
National Research University Higher School of Economics, Kochnovskiy Proezd 3, 125319 Moscow, Russia
e-mail: iamakarov@hse.ru; revan1986@mail.ru

O. Bulanov
e-mail: oleggl500@gmail.com

L.E. Zhukov
e-mail: lzhukov@hse.ru

V.A. Kalyagin et al. (eds.), *Models, Algorithms, and Technologies for Network Analysis*, Springer Proceedings in Mathematics & Statistics 197,
DOI 10.1007/978-3-319-56829-4_18

while in [4] authors gave a representation of finance network analysis. In [5, 6] the authors studied correlation between citation indexes and centrality measures in a co-authorship network and in [7] predicted citation indexes from centrality metrics. There are also exist numeral publication studying general features of co-authorship networks in various science fields [1, 8, 9]. The methods and applications of network analysis were described in [10].

In this paper, we present a co-authorship network based on papers co-authored by researchers from the National Research University Higher School of Economics (HSE). HSE authors often have publications in collaboration with non-HSE authors, so it is necessary to include such authors to the network but the recommender system gives recommendations only among HSE researchers. Non-HSE authors were added in order to calculate network nodes metrics more precisely.

We started by taking a relational database of all publications written by NRU HSE authors. We cleaned the database by removing duplicate records and inconsistencies, unified author identifiers and filled in missing data. An undirected co-authorship graph was constructed with authors as graph nodes and edges containing lists of jointly published papers. We added all publication attributes from the university portal and subject areas and categories from Scopus Journal Ranking [11, 12]. Publication quality was taken from its quartile in SJR ranking for the publication year, computed as maximal (or average) over different categories per journal. Information about author's administrative units and declared author research interests was also included as node features.

The co-authorship graph can help to answer a variety of research questions about collaboration patterns: distribution of number of papers written by an author and distribution of number of collaborators, density of graph communities, dependency on research area and administrative units, evolution of collaborations over time, etc. We use the graph to power a co-author recommender system. The system gives recommendations of authors that have interests similar to the chosen author or whose co-authorship pattern is similar to that of the author. More specifically, for a selected author the system generates a ranked list of authors whose papers could be relevant to him, and authors themselves could be good candidates for collaboration.

2 Author Similarity Score

Let us consider problem of finding authors with similar interests to a selected author. We formulate this recommendation problem as a problem of link prediction in the network and use similarity between network nodes for prediction. The comparative analysis of network similarity metrics is provided in [13].

All similarity metrics can be divided into two types. The first four metrics from the Table 1 are standard similarity metrics described in [13]. Since nodes of the co-authorship network represent known authors from HSE, one can also define additional content-based features for similarity metrics between authors'. We used the following content-based features: the number of fields of journals, the number of

Table 1 Similarity metrics

Similarity metric	Definition
Common neighbors	$sim(v_i, v_j) = \|N(v_i) \cap N(v_j)\|$ where $N(v)$ is number of neighbours and v, v_i, v_j are nodes
Jaccards coefficient	$sim(v_i, v_j) = \frac{\|N(v_i) \cap . N(v_j)\|}{\|N(v_i) \cup N(v_j)\|}$
Adamic/Adar	$sim(v_i, v_j) = \sum_{v \in N(v_i) \cap N(v_j)} \frac{1}{\ln \|N(v)\|}$
Graph distance	Length of the shortest path between v_i and v_j
Cosine	$sim(a, b) = \frac{(a,b)}{\|\|a\|\| \cdot \|\|b\|\|}$
Interests	Normalized number of common journal SJR areas

papers, the number of papers in journals of high quartiles, the number of papers during past 3 years, etc., local clustering coefficients, degree, betweenness and eigenvector centrality metrics, position and seniority. We calculated cosine similarity for a vector consisting of normalized values of the feature parameters and "interests" metric for the journals where papers where published.

3 Choosing Subgraphs for a Training Set

HSE co-authorship network contains nodes that represent authors from different fields of study and departments. In [1] Newman showed that researchers from different scientific areas form new connections differently. So we first created overlapping groups of authors from similar affiliation and scientific interests.

We start with forming department subgraph, defined by unit staff membership for HSE co-authors. We then construct a feature vector for a department consisting of over 30 different descriptive statistics of the department subgraph, quantities of publications and normalized publications activity of the researchers with respect to different time intervals and quartiles. We considered two departments *similar*, if the norm of the difference between their feature vectors is less than the median of the distances between the pairs of the feature vectors for all departments.

We select candidates from each department by the following procedure. Initially, all the authors from the department, as well as the authors from similar departments are considered as candidates. Every author with the same areas of journals as those of the selected author's department is also added as a candidate. We used five methods of community detection on the co-authorship network, such as label propagation [14], fastgreedy [15], louvain [16], walktrap [17], infomap [18], and created five candidate sets by unifying the previous set with all the found communities containing authors from the previous set. Finally, all non-HSE authors were removed from these sets. The Euler diagram of the obtained groups is shown at the Fig. 1.

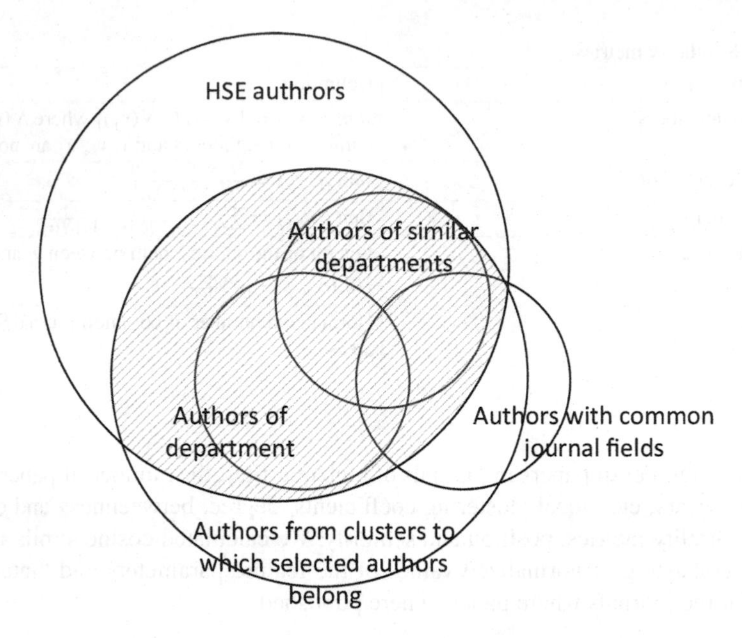

Fig. 1 Euler diagram of group of authors among which recommendations will be given

4 Recommender System

We used linear regression on normalized feature vectors to predict new links. We applied lasso regularization and choose high regularization parameter. Equation 1 shows a lasso regression, where X is matrix of similarity metrics values, Y is a vector indicating links' presences, λ is a regularization parameter and N is a number of similarity metrics.

$$\begin{cases} \frac{1}{N}(Y - X\theta)^T(Y - X\theta) + \lambda||\theta||_1 \longrightarrow \min_\theta \\ ||\theta|| < \lambda \end{cases} \tag{1}$$

Let us describe the process of constructing the recommender system similar to [19]. For a given researcher, we form corresponding subgraphs for training sets from the previous section and choose only that contained our researcher. We construct linear regression model for each of the groups, taking as positive examples links in the chosen subgraph, and the same number of negative examples as missing links in the same subgraph. For a fixed group, we choose one community detection method with the highest precision among five corresponding to different clustering methods (see Algorithm 1).

Algorithm 1: Algorithm of constructing a recommender system

Data: N - co-authrship network
Result: $\{\theta\}$ - regression coefficients for each group, $\{G\}$ - groups
begin
  $\{G\} \longleftarrow$ all groups, five for each department
  for $G \in \{G\}$ **do**
    $X_{train} \longleftarrow$ features of links and the same number n of features of nonexistence links
    $Y_{train} \longleftarrow (1 \ldots 1, 0 \ldots 0)$ with n units and n zeros
    $(\theta, \lambda) \longleftarrow$ calculate regression coefficients fitting λ for highest precision
  $\{G\} \longleftarrow$ select one group for each department

Table 2 Similarity metrics

	Precision	Recall	Accuracy	F1-measure	AUC
Train data	0.916251	0.991259	0.9467785	0.949979	0.990913
Test data	0.901435	0.867798	0.8733743	0.870438	0.923516

After we compute all linear models for each of the groups, we can describe the scheme of making a recommendation. At first, we choose a group with the highest precision corresponding to one of the departments, which the selected author belongs to. A group should be fixed because an author may belong to several departments simultaneously. Second, we take a normalized vector of predictions from linear regression. We provide a ranked list of recommendation ordered by the predicted values that are greater than 0.5.

5 Results

We predicted links between those authors that have written from $k = 1$ to $k = 5$ papers together and obtained a series of, so-called, "strong" subgraphs to use in cross-validation. For all the pairs of k_1/k_2-subgraphs, we calculated the predictions for the "stronger" subgraph. For each group, we build two subgraphs induced by authors from a group in the corresponding stronger subgraph. We find the difference of the link sets for k_1 and k_2 stronger subgraphs ($k_1 < k_2$). If the difference is not empty, we prepare the test sample as a set of links from the difference and the same number of missing links from the links difference with features taken from the stronger subgraph, otherwise, we change a group. For all the groups, we calculate average error rates for test and train sets over all pairs of thresholds values of k (see Table 2). The area under the rock curve (AUC) and F1-measure are high, therefore, normalized lasso regression is sufficient for binary classification.

6 Conclusion

We developed a recommender system based on HSE co-authorship network. The recommender system demonstrates promising results on predicting new collaborations between existing authors and can fasten the process of finding collaborators and relevant research papers. The recommendation system can be also used for new authors, who do not have any connections to HSE community. A further analysis of the co-authorship network may help stating university policy to support novice researchers and increase their publishing activity and estimate collaboration between the university departments. Though tested on HSE co-authorship network, the approach can be easily applied to other networks.

Acknowledgements I. Makarov was supported within the framework of the Basic Research Program at National Research University Higher School of Economics and within the framework of a subsidy by the Russian Academic Excellence Project '5-100'

References

1. Newman, M.E.J.: Coauthorship networks and patterns of scientific collaboration. PNAS **101**(suppl 1), 5200–05205 (2004)
2. Newman, M.E.J.: Who is the best connected scientist? a study of scientific coauthorship networks. In: Complex Networks, LNPh, pp. 337–370. Springer, Heidelberg (2000)
3. Morel, K.M., Serruya, S.J., Penna, G.O., Guimaraes, R.: Co-authorship network analysis: a powerful tool for strategic planning of research, development and capacity building programs on neglected diseases. PLOS Negl. Trop. Dis. **3**(8), 1–7 (2009)
4. Cetorelli, N., Peristiani, S.: Prestigious stock exchanges: a network analysis of international financial centers. J. Bank. Financ. **37**(5), 1543–1551 (2013)
5. Li, E.Y., Liaoa, C.H., Yenb, H.R.: Co-authorship networks and research impact: a social capital perspective. Res. Polic. **42**(9), 1515–1530 (2013)
6. Yan, E., Ding, Y.: Applying centrality measures to impact analysis: a coauthorship network analysis. J. Am. Soc. Inf. Sci. Technol. **60**(10), 2107–2118 (2009)
7. Sarigl, E., Pfitzner, R., Scholtes, I., Garas, A., Schweitzer, F.: Predicting scientific success based on coauthorship networks. EPJ Data Sci. **3**(1), 9p (2014)
8. Velden, T., Lagoze, C.: Patterns of collaboration in co-authorship networks in chemistry-mesoscopic analysis and interpretation, ISI - 2009, vol. 2, 12p. (2009)
9. Zervas, P., Tsitmidell, A., Sampson, D.G., Chen, N.S., Kinshuk.: Studying research collaboration patterns via co-authorship analysis in the field of tel: the case of ETS journal. J. Educ. Technol. Soc. **17**(4), 1–16 (2014)
10. Wasserman, S., Faust, F.: Social Network Analysis Methods and Applications. Cambridge University Press, Cambridge (1994)
11. Gonzlez-Pereira, B., Guerrero-Bote, V.P., Moya-Anegn, F.: A new approach to the metric of journals scientific prestige: the SJR indicator. J. Inf. **4**(3), 379–391 (2010)
12. Guerrero-Bote, V.P., Moya-Anegn, F.: A further step forward in measuring journals scientific prestige: the SJR2 indicator. J. Inf. **6**(4), 674–688 (2012)
13. Liben-Nowell, D., Kleinberg, J.: The link prediction problem for social networks. J. Am. Soc. Inf. Sci. Technol. **58**(7), 1019–1031 (2007)
14. Kumara, S., Raghavan, U.N., Albert, R.: Near linear time algorithm to detect community structures in large-scale networks. Phys. Rev. E **76**(3), 12p (2007)

15. Moore, C., Clauset, A., Newman, M.E.J.: Finding community structure in very large networks. Phys. Rev. E **70**, 6p (2004)
16. Blondel, V.D., Guillaume, J., Lambiotte, R., Lefebvre, E.: Fast unfolding of communities in large networks. J. Stat. Mech. **2008**(10), 12p (2008)
17. Latapy, M., Pons, P.: Computing communities in large networks using random walks. In: Computer and Information Sciences - ISCIS 2005. LNCS, vol. 3733, pp. 284–293 (2005)
18. Bergstrom, C.T., Rosvall, M.: Maps of random walks on complex networks reveal community structure. PNAS **105**(4), 1118–1123 (2008)
19. Beel, J., Langer, S., Genzmehr, M., Gipp, B., Breitinger, C., Nurnberger, A.: Research paper recommender system evaluation: a quantitative literature survey. ACM RepSys, 15–22 (2013)

Network Studies in Russia: From Articles to the Structure of a Research Community

Daria Maltseva and Ilia Karpov

Abstract Our research focuses on the structure of a research community of Russian scientists involved in network studies, which is studied by means of analysis of articles published in Russian-language journals. The direction of network studies in Russia is quite new form of research methodology—however, in recent years we can observe the growing number of scientists working at this direction and institutionalized forms of their cooperation. Studying the structure of these researchers' community is important for the fields development. This paper is the first report on the research, that is why it focuses on methodological issues. It covers the description of method of citation (reference) analysis that we use and the process of data collection from eLibrary.ru resource, as well as presents some brief overview of collected data (based on analysis of 8 000 papers). It is concluded by representation of future steps of the research.

1 Introduction

The development of a certain science discipline in many respects depends not only on institutional context—the official approvement of discipline and presence of organizations engaged in certain type of research—but on the structure of informal (implicit) social and communicational structures of researchers as well. Such system of relations between researchers was developed in the sociology of science by Diana Crane in 1972 building on Derek de Solla Price's work on citation networks and called "invisible college", meaning the informal (implicit) social and communicational structures of researchers, who refer to each other in their publications without being linked by formal organizational ties. The usage of this notion made it possible to find out the presence of some new fields and disciplines.

D. Maltseva (✉) · I. Karpov
International Laboratory for Applied Network Research,
National Research University Higher School of Economics, Moscow, Russia
e-mail: d_malceva@mail.ru

I. Karpov
e-mail: karpovilia@gmail.com

V.A. Kalyagin et al. (eds.), *Models, Algorithms, and Technologies for Network Analysis*, Springer Proceedings in Mathematics & Statistics 197,
DOI 10.1007/978-3-319-56829-4_19

Network study as a form of research methodology, which operates with the notion of networks for studying different social phenomena, was one of the fields, which was recognized as a separate discipline in 1970s in Western sociology. The discipline called "Social network analysis" (SNA) was characterized both by institutionalized forms—journals, conferences, knowledge transfer centers, and educational programs—and the presence of its own professional community (informal "invisible college") [15].

The exclusion of Russia from the context of social sciences, which was typical for the Soviet period, has further led to certain lags in some areas, including network studies. However, during recent years we can observe the growing interest to this new form of a research methodology—the usage of social network analysis technics becomes evident and "fashionable" in Russian scientific space (which can be seen by the increase of journal publications), the appearance of scientists who nominate themselves as "network researchers" and the development of institutionalized forms of their cooperation (e.g., research sections at universities and organizations, laboratories).

However, there is no information on the characteristics of the community of network scientists in Russian-language space, language yet—who are the main drivers of the field's development, if they consider themselves as cooperators, representing "invisible college", or see each other as competitors, if they interact with each other or mostly prefer some "significant others". The literature review of the works of Russian scientists involved into the field of network studies in the sociology shows that different authors regard different—but foreign—scientists as "founding fathers", whose works are important for the filed establishment and development, such as B. Wellman, S. Wasserman, K. Faust, L. Freeman [10], B. Wellman, S. Berkowitz, M. Granovetter, H. White [26], R. Emerson, K. Cook, J. Coleman [18], R. Emerson, K. Cook, L. Molm, M. Emirbayer, anthropologists [7], B. Latour, J. Law, M. Callon [17, 31].

This situation makes it important for the current state of field's development to study the structure of a research community of scientists involved into network studies in Russia—who are these main drivers, how they relate to each other, and at what research teams—Russian or foreign—they are mainly focused. We propose to build the structure of this research community basing on the quantitative method of citation (reference) analysis of articles on "network" topics published in Russian journals in different disciplines, which are provided by the largest electronic library of scientific periodicals in Russian eLibrary.ru.

As many articles presenting the results of citation analysis often do not provide enough information on their methodology to reproduce the study or rationale for methodological decisions [14], in the present article we would like to cover some methodological issues concerning the proceeding study and present the overview of the method of citation (reference) analysis as a tool for studying scientific fields and describe the process of data collection in detail. We propose that such description can be interesting for the scientists who do not have an experience of working with Russian-speaking platform of scientific measuring. Providing the brief information on already collected data on articles on network topics, we conclude with the description of future steps of the study.

2 Citation Analysis as a Tool for Studying Science

In the first section we present some general information on the method of citation analysis as a special tool used for studying scientific fields. As we are mostly interested in sociology and social network analysis, we provide some examples of the previous studies done in these disciplines using observed tool.

2.1 Citation Analysis as a Method

Citation analysis is a method used in the field of informetrics (more precisely, its subareas bibliometrics and scientometrics) for the study of different forms of social interaction networks, including authors' citation networks, co-citation networks, collaboration structures, and other (look [2] for a detailed review). Citation analysis was established as an instrument of managerial control of modern science, which was institutionalized in the middle of twentieth century and changed the practice of references from concrete names to precisely dated texts [21]. The first well-known usage of a citation analysis is associated with the name of the chemist Eugene Garfield, who developed the first and revolutionary citation index Science Citation Index, SCI, in 1955, as a representative of the Institute for Scientific Information (now Thomson Reuters) [6, 22].

Even though Garfield's works were innovative for the filed, there were some other authors who could be assumed as pioneers of this methodological approach. The first paper that can be considered as a citation analysis was published in 1927 by Gross and Gross, who studied the references found during 1 year's issues of the Journal of the American Chemical Society. According to Casey and McMillan [4], even though the term bibliometrics is dating only to the late 1960s, the field itself has roots reaching back at least 80 years to the work of Lotka published in 1926. Garfield himself, in his highly cited article [9], enumerates names of others citation analysis pioneers such as Bradford, Allen, Cross and Woodford, Hooker, Henkle, Fussler and Brown. De Bells [6] names John Desmond Bernal and Derek John De Solla Price as "philosophical founders of bibliometrics". Other important names, due to de Bells, are sociologist Robert Merton and chemist and historian of science, leading researcher with Garfield's former company Henry Small. From its beginning, now citation analysis has grown into a developed field. During the recent decades, there was a substantial literature on citations [30]. In 1980, Hjerrpe published a review with more than 2 000 items of research on this and related topics [22]. It is appropriate to note that in 2015 in Web of Science data base there were more than 2 500 publications found by the query of citation analysis, starting from the Garfields 1972 work as the most cited article [9].

Citation itself can be understood as a complex phenomenon, which considers interaction between networks of authors and texts, that is why it indicates not only cognitive, but also social contexts of a knowledge claim [21]. However, it is important to clarify the differences between citation and *reference*. Although in practice many researchers do not divide these terms and use them interchangeably, each of them represents a different entity in the citing or cited perspective. The reference is made within a citing document and represents an acknowledgement of another study (and can be measured by the number of items in its bibliography as endnotes, footnotes, etc.), while a citation represents the acknowledgement received by the cited document from other publications (and can be seen in citation index) [2, 28].

Talking about citations in meta-level, citations can be viewed as explanans (something explaining something else) and *explanandum* (something to be explained). While a lot of interest is usually given to the first notion when citations explain research impact and value, the question of what is a citation should be brought up for the discussion in order to better understand what is certainly measured [21]. Citation analysis often starts with the assumption that references are indicators of influences on other scientists work. According to normative theory, norms of science suppose that authors cite works that they found useful for their own research, and it is assumed that they abide these norms, citing some authors and thus giving credits where they are due (as citing A by B means that A's works influenced B's thinking) [22]. Citation is as well considered as an indicator of reward in the science system in evaluation studies for science policy purposes (Martin and Irvine; Moed; Luukkonen; Merton; Latour and Woolgar in [21]), symbolic payments of intellectual debt, or *representation of trust* in virtual environments, that makes citation indexes to be "recommender systems" for other scientists [2]. Some practical reasons of citation-making process were also enumerated by Garfield (in [28]). In other scientific traditions, citation was also seen as a *function in scientific communication among texts* (Cronin, in [21]). Much attention was paid to the perfunctory and *rhetorical functions of citations* within the scientific community by B. Latour. Some investment into the citation analysis' theoretical legitimation (the theory of what is being analyzed) was made in the field of Science and Technology Studies (STS) itself, which included formalized measurement of citation analysis into empirical studies in 1980s. Nevertheless, theoretical and methodological reflection is still needed as there is a need of "translation" qualitative side of STS and merging it with formal approach [21].

During its history, citation analysis has proved to be a well-established tool for different aims of science analysis, including measurement of research impact and value [5, 14, 23]. However, besides all the discoveries, there was a substantial critique of this method [22, 28, 30, 34]. The scientists came up to the idea that it is not advisable to use citation analysis as a single and absolute criterion for judging the importance of a publication. Giving objective information regarding an individual, research group, journal or higher education institution, this method should be supplemented by other kinds of analysis, including qualitative approach (qualitative review, peer assessment, studying the authors behavior, characteristics of documents cited and not cited) [3, 28, 30, 34].

2.2 Citation Analysis as a Tool for Studying Scientific Fields

Methods of citation and co-citation analysis were used for studying of different aspects of scientific communication: coauthorship networks as complex systems [1, 8], dynamic aspects of collaboration networks [24], international collaboration as a self-organizing network based on the principle of preferential attachment [32], social ties, co-citations and inter-citations of Globenet, offline and online collaboration [35]. More examples of the studies are presented in [2].

One of the first examples of bibliometric tools usage for studying Social Network Analysis field was conducted by Hummon and Carley [15], who analyzed first volumes of Social Network Journal and other articles and have found that there is an invisible college in the growing SNA field. As the journal so specifically displays the people involved into network discipline it is quite often used as a looking glass on the social networks community [11]. Basing on the articles in the same journal, Lewis compared two types of social networks—formal collaborative relationships represented by coauthorship (who publishes papers with whom) and the informal collaborative relationships represented by acknowledgment (who thanks whom in published papers) in the scientific community of social network analysts [20]. Analyzing the Sociological abstracts data base, Otte and Rousseau [25] studied the underlying collaborative relationships between authors, built coauthorship network, pointed out central players of the field, and showed connections between SNA and other subfields (especially Information sciences).

In Russian scientific space studies, bibliometrics methods are not so much developed, even though in 1980s the technique of co-citing was developed by I. Marshakova, in parallel with G. Small [36]. However, there are also some examples of citation analysis usage for the studying of scientific fields. Cognitive structures of Russian Sociology and Ethnology by the method of co-citation analysis were studied by B. Winer, K. Divisenko and M. Safronova [27, 36], who tested different methods of the cognitively closed groups detection. Among other methods, citation analysis was used in the study of intelligent landscape and social structure of the local academic community (the case of St. Petersburg) [29]. The authors found three groups among Russian sociologists (West-side, East-side and Transition zone), who tend to see (in the meaning of citing) the representatives of their own groups, while the authors from other groups stay almost "invisible" for them.

It is important to note that most of the studies regard citation in the first sense—as the acknowledgement that the author gets from other scientists, but not a credit that he or she gives to them,—that is why what is being used is the method of citation analysis. In our project, we understand the citation in the second meaning—as an acknowledgement of another study to the current study—and propose to use the method that can be called reference analysis, where reference means a tie between the author (writer) of article and the author whom he or she cites in publication. Basing on authors of articles and authors from their bibliography lists allows us to build networks of relations between different groups of authors and study the structure of a community of researchers involved into network studies in Russian scientific space.

3 Data and Methodology

In this section we present some practical information concerning our data source and process of data collection and preprocessing (such as author disambiguation and paper classification) and discuss some problems associated with these procedures.

3.1 Data Source

The data source that we use is the electronic library of scientific periodicals in Russian called eLibrary.ru—a leading electronic library of science periodicals in Russian, which contains more than 3 900 Russian-language and 4 000 foreign scientific journals, abstracts of nearly 20 000 journals and the descriptions of 1.5 million of Russian and foreign dissertation thesis, and has 1.1 million individual users and 2 200 organizations registered. The base is integrated with the Russian science citation index (RSCI)—a national information–analytical system which accumulates more than 6 million of publications of Russian authors in more than 4 500 Russian journals, as well as information on citing of these publications. Even though the system is based on indexed articles in Russian scientific journals, in recent years other types of scientific publications were included into the base—such as reports on conferences, monographs, tutorials, patents, and dissertation thesis. The chronological coverage of the system comes from 2005, but for many resources the depth of archives is deeper. In sum, eLibrary resource not only gives the support of scientists by the bibliographic information, but provides a tool for assessing the effectiveness of science and research organizations (more than 11.000) and scientists (more than 600.000), as well as scientific journals.

Unfortunately, when we go from the level of description of the resource to its practical usage, some problems associated with data collection appear. First of all, the resource does not offer any procedures for mass data downloading, as some scientific aggregators as Web of Science allow. The data collection process needs manual collection, which is impossible in the situation of a large amount of data, or the special crawling techniques.

3.2 Data Collection and Preprocessing

For each article, the eLibrary base contains information on publisher's imprint, paper's title, authors, their affiliations, keywords and disciplines, abstracts, and what is the most important in the terms of the current study—lists of bibliography (the references) and lists of other eLibrary papers—that cited the initial paper. That is why our data contains two parts. Data base (1) contains all the information on articles journal's name, discipline, year of publication, author's organization, keywords, annotation, scientometric indexes, etc. Data base (2) contains information on references—main data that contains authors and lists of bibliography.

The method of data collection that we used is based on expanding publication graph using two strategies:

1. Expansion strategy—a set of methods to increase the number of relevant papers (increases recall):
   a. Keyword search—we formed a list of 48 Russian and English keywords that we consider relevant to the domain of network research and collected all publications. Given the search query, eLibrary engine returns all articles that contain the search query in any field, including title, keyword, or annotation. Keyword list contains such keywords as *network analysis*, *relational sociology*, *actor-network theory*, *graph of a network*, etc. Having a list of relevant articles, obtained after filtering, we generated the distribution of articles for each keyword and selected the keywords that appeared more than the median value. Afterward, we repeated step a. with newly obtained keywords.
   b. Author search—if author had more than three publications in the domain of network research, we collected all his publications and filtered them with our keyword classifier;
   c. Citation search—we took all papers that cited the article from the network research domain.
2. Filter strategy—a set of methods to remove irrelevant papers, found during the expansion strategy (increases precision). It often happens that relevant keywords are used in a different meaning or separated by other words. Besides, many authors publish papers in multiple research areas or cite articles from other domain. In all these cases we often collect irrelevant papers that must be filtered. Traditionally, document classification task is made by machine learning classification, but by virtue of the fact that we have no annotated collection of documents, we used two-step strategy:
   a. Cluster documents using keywords and annotation text and manually mark relevant clusters.
   b. Convert chosen clusters to binary classes (relevant/irrelevant) and make a binary classification of the entire array. Filter strategy problem is described in more depth in Paper classification section.

Proposed method is very similar to the shark-search approach [13], but the relevance of each cluster is annotated manually. The following limitations should be taken into account:

1. Clustering forms lexically similar groups, but cluster center and periphery may contain very different articles;
2. Having a list of totally unconnected fields, we need at least one “seed” keyword in each field to make expansion strategy work;
3. Papers that do not contain common keywords can be lost during the filtering step.

There is no standard evaluation task for the proposed method as we cannot obtain recall for the 20 m collection, but we have made evaluation for our filtering strategy on the Cora Research Paper Classification task.[1] We compared each rubric with the most relevant cluster and obtained F1-score = 0.64 for automated and 0.92 for semi-manual clustering where F1-score is the mean score among all rubrics.

Brief description of typical parsing problems associated with collected fields is provided below:

- Author's name and surname of each author. We excluded papers with more than 10 authors because we consider their interconnectivity to be very weak and the problem of combinatorial explosion at the step of artifacts generation. We also cleaned up special eLibrary markup, such as editor or translator, and special author affiliations–organizations related to the certain author.
- Keywords—mentioned by author and splitted by comma. We excluded articles with one keyword consisting of more than five words.
- Language of the article. Article, which may differ from the abstract and keywords language. We used external language detection tool based on sequences of characters information.
- Abstract—short text, describing the article, which may be written in multiple languages (for example in Russian, French and English in paper 11897467 [2]). In this case we kept only one language (priority is Russian, English, other languages).
- Citations—list of papers, cited by this paper. Citation list is unstructured and very dirty, so we extracted only surnames information.

As there are different strategies of author counting in the literature (see [2, 22]) we decided to use all authors instead of first authors counting of citations and to include self-citing into the collection (as it is entirely appropriate for scientists to build on their own previous studies).

Such way of data collection allowed us to get not only the information on ties of "citing" ("referencing") type (1), but also the data on coauthorship in the article (2) and coauthorship in the citing article (3)—when there was more than one author in citing and cited articles. Also the data on ties between authors and "artifacts" was collected, where the latter were author's affiliation (organization) (4) and concepts that he or she uses in the works (5). Thus, the collected data potentially allows us to analyze five types of ties, and conduct more complex study in future, including semantic, citation, co-citation, analysis of coauthorship and affiliation networks, analysis of ties between authors and concepts, as well as work on such methodological issues as comparison of methods of articles sampling, as shown in Fig. 1.

[1] Download at http://sites.google.com/site/semanticbasedregularization/home/software/experiments_on_cora.

[2] Example is available by the link: http://elibrary.ru/item.asp?id=11897467.

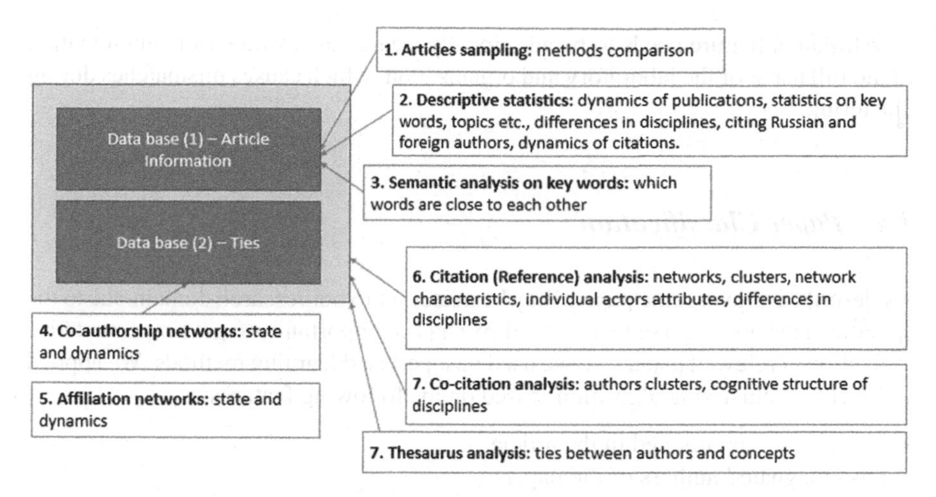

Fig. 1 Possible directions of analysis

3.3 Author Disambiguation

As many researchers work in the same field, papers made by different researchers with the same name should be searched out during the collection process. The process of resolving conflicts that arise when a potential name is ambiguous is called disambiguation. In some cases disambiguation is provided by eLibrary itself, and the authors are marked with special hyperlink and identifier, that is unique in the whole data base. In other cases, we have to solve two problems:

- How many unique authors with the same name are represented in the collection?
- How to classify each ambiguity to one of the existing classes?

Additional problem appears when we have the same author published in English and Russian languages. The core problem of matching Cyrillic and Latin names is that usually author tends to save phonetics of the word that does not match any transliteration rules. Our match was based on idea that most Russian authors obtain at least one English article with correct English name that can be used as correct transliteration. English authors were transcribed into Russian according to the GOST 7.79 2000 standart. We searched for all possible surnames with Levenshtein distance ≤ 3 [19] and manually validated them. Each author was trimmed to surname and initials, and we used hierarchical agglomerative clustering [16] to label each author-in-the-article pair to the certain cluster, based on the following features:

1. Keywords—mentioned in the article;
2. Coauthors—surnames of the coauthors for the certain author;
3. Affiliations—organizations, related to the ambiguous author;
4. Date of the publication—year of publication (is very weak feature as we have a contemporary field);
5. List of citations—surnames of the cited authors.

Affiliation feature needs to be additionally cleared as it varies from abbreviation to the full name of the laboratory and organization, which causes mismatches during clustering.

3.4 Paper Classification

As described above, some papers may be irrelevant to our research domain due to the specific of the proposed search method. We had no opportunity to get a representative collection of relevant papers, so we used unsupervised learning methods. We applied BIRCH [37] clustering algorithm, based on the following features:

- Keywords— mentioned in the article;
- Disambiguated authors of the paper.

Clustering hyperparameters are as follows: number of clusters 64–255, number of top terms 50 000, term weighting method

$$W_i = \sigma^2 \left(\frac{TF_i}{IDF_i} \right)$$

where i is term id, TF_i—i-th is the term frequency, frequency of the given term in the document, IDF_i is the inverted document frequency, the measure of fraction of the documents that contain the i-th term in the whole collection, distance metric—Ward's method [33]. Overall collection process consisted of four consecutive phases: $Search\ phase\ 1 \rightarrow Filter\ phase\ 1 \rightarrow Search\ phase\ 2 \rightarrow Filter\ phase\ 2$. Resulting dataset is described in Table 1 below. The resulting number of articles is composed of 8 260.

Table 1 Resulting dataset statistics

	Search phase 1	Filter phase 1	Search phase 2	Filter phase 2
Number of articles	220 657	5 836	442 524	8 260
Added by title search	220 657	–	107 208	–
Added by author search	0	–	56 932	–
Added by citation search	0	–	181 867	–
Filtered by article type	–	121 880	–	234 114
Filtered by clustering	–	90 941	–	200 150

4 Results

In this section we will briefly provide the overview on the data on articles (Data base 1). We analyzed articles from the resulting data collection, which is more than 8 000 papers. Main information on this sample is shown in Table 2 (amount, min. and max. citing, the earliest and the latest year of publication).

The growing number of publications collected is shown in Fig. 2, with the first article published in 1988. During recent years we can see the growth of interest to the network topics—in last 5 years, from 2010 to 2015, the number of articles increased almost in four times. However, the low annual amounts for previous years might be associated with the quality of the data base itself.

Talking about types of the articles, most of the presented publications are articles in journals (scientific articles), which form 67% of the entire sample (Table 5). Second and third places are taken by PhD thesis (11%) and articles in the conference proceedings (9%). Other types of publications can be met less frequently.

Table 2 Main characteristics of the sample

Number of articles	8 260	100%
Mean citing	2.13	–
Min citing	0	–
Max citing	303	–
The earliest year of publication	1988	–
The latest year of publication	2016	–

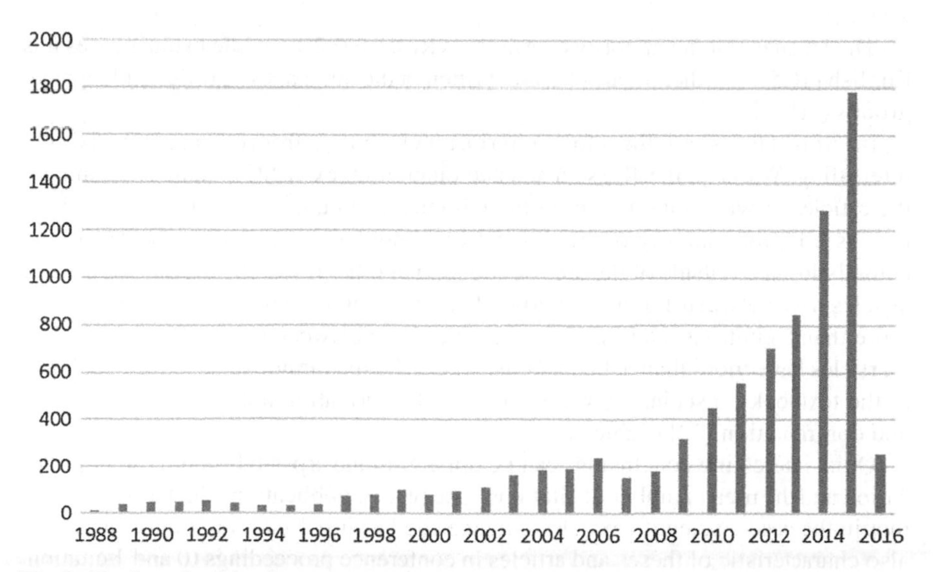

Fig. 2 Number of articles, by years

Table 3 Language of articles

	Frequencies	% by column
Russian	6 807	82
English	1 362	16.5
Other	58	0.7
Not defined	33	0.4
Total	8 260	100

Table 4 Number of citations

		Frequencies	%	valid %
Valid	0	5 132	62.1	62.5
	1–5	2 360	28.6	28.7
	6–20	579	7.0	7.1
	21–50	104	1.3	1.3
	51–100	26	0.3	0.3
	101–150	5	0.1	0.1
	151–500	4	0.05	0.05
	Total	8 210	99.4	100.0
Missing	System—missing	50	0.6	
Total		8 260	100.0	

The number of PhD thesis is quite high–948 dissertations, with the first published in 1997.

The language of the majority of articles is Russian (82%), while in other cases it is English (16.5%). Other 58 articles are written in the languages of different language groups (Table 3).

In terms of this study, the information on articles citing (from other authors) is quite interesting. We used the Russian science citation index (RSCI) values to compare the articles. It was found that even though the mean number of citing in the whole base is 2.13, the majority of articles (62%) do not have any citations in RSCI, i.e., actually located outside of the area of attention of other researchers. Another 29% of articles have between 1 and 5 citations. Thus, 91% of articles in general do not have more than 5 citations, and just 7% of articles have between 6 and 20 citations. Just 9 articles have more than 100 citations, where the maximum value—303—belongs to the textbook on social networks, models of information influence, management, and confrontation [12] (Table 4).

Quite interesting conclusions can be done from the cross-tables of type of publication and the mean number of citations. The type of publication which is most often met in the base—scientific articles—in average has just 1 citation. Small values are also characteristic of theses and articles in conference proceedings (0 and 1 citation, respectively). The highest amounts of citations are typical for monographs (2% of

Table 5 Citation in RSCI by types of citations

	Freq.	% by column	Mean	Min	Max
Article in journal—scientific article	5 521	67	1	0	86
Thesis	948	11	3	0	71
Article in conference proceedings	728	9	01	0	29
Article in journal	282	3	2	0	59
Monograph	195	2	18	0	292
Articles in the digest of articles	115	1.39	1	0	44
Article in journal—review article	96	1.16	2	0	27
Article in journal—conference materials	68	0.82	0	0	4
Tutorial	66	0.8	16	0	303
Abstract at the conference	49	0.59	0	0	3
Thesis abstract	42	0.51	4	0	28
Article in journal—other	39	0.47	0	0	3
Article in the journal—abstract	30	0.36	0	0	1
Article in the journal—a short message	12	0.15	1	0	7
Methodological guidelines	10	0.12	3	0	20
Digest of articles	10	0.12	1	0	5
Chapter in a book	9	0.11	0	0	3
Article in the journal—review	7	0.08	0	0	1
Dictionary or reference book	6	0.07	3	0	11
Article in the journal—editorial note	4	0.05	1	0	2
Article in the journal—scientific report	3	0.04	2	0	5
Report on research work	2	0.02	0	0	0
Article in the journal—correspondence	2	0.02	0	0	0
Article in the journal—personality	1	0.01	0	0	0
Deposited manuscript	1	0.01	0	0	0
Article in an open archive	1	0.01	0	0	0
Brochure	1	0.01	7	7	7
Other	12	0.15	0	0	0
Total	8 260	100	2	0	303

sample, mean value 18) and tutorial (1% of sample, mean value 16) (Table 5). Among monographs and tutorial just 29% and 18% of articles, respectively, have the number of citations between 1 and 5; while 22% and 26% of them, respectively, between 5 and 20, and 17% and 11%, respectively, between 21 and 50. This outcome correlates well with the results of foreign studies, in which it has been proved that there are the differences in citation practices between books and journals, with the greater emphasis given to the books [30].

Table 6 Network of keywords measure parameters

Vertices	143
Unique edges	1302
Edges with duplicates	2138
Total edges	3440
Avg degree	33
Maximum geodesic distance (Diameter)	4
Average geodesic distance	1.87
Graph density	0.23
Connected components	1

We provide these results with the thematic map of the data, which was constructed by network analysis of the main keywords used in the collected articles (Fig. 3). Extracted keywords which were met in the dataset more than 30 times (min. value = 30, max. value = 1888) were used for building this network. In the network, which was constructed by NodeXL software (by Harel-Koren Fast Multiscale layout), the scale of the nodes corresponds to the frequency the keyword was met in the dataset, the color of nodes corresponds to their degree value, edges are weighted according to the number of times two keywords were met in the same article, and the edges are filtered (min. value = 4 and max. value = 801). The information on network is provided in Table 6.

Figure 3 shows map of different groups of keywords, which are more likely to meet in the same article and thus are close to each other by topics. In the network center, there are words more frequently used in the dataset—social network and Internet, both in Russian and English. These words are connected to a large group of keywords above them associated with Internet communication technologies such as blogosphere, social networking, and some sites like VKontakte, Facebook, Twitter, Web 2.0, social media, and media itself. Quite large group is associated with Internet marketing promotion, content, site, SMM. Close to them we can also see the groups of words associated with Internet communication, self-presentation, and Internet addition, as well as information security and personal data.

Main keywords "Social Network" and "Internet" are also associated with information technologies, communication, and information society. Another large group (at the bottom) consists of the words associated with society—social capital, human capital, trust, civil society, and social inequality. On the right side, we can see a large group of words associated with education—educational organizations, educational programs, professional standard, modernization of education, and academic mobility. There is another large group on the left side, which consists of the vocabulary from humanities—Sociology, Philosophy, Social philosophy, Political science, Economics, International relationships, Conflictology—such as Social and economic institutions, Social processes, Social structure, Social community, State, Power, etc. Thus, we can see that the thematic map of our dataset covers different aspects of

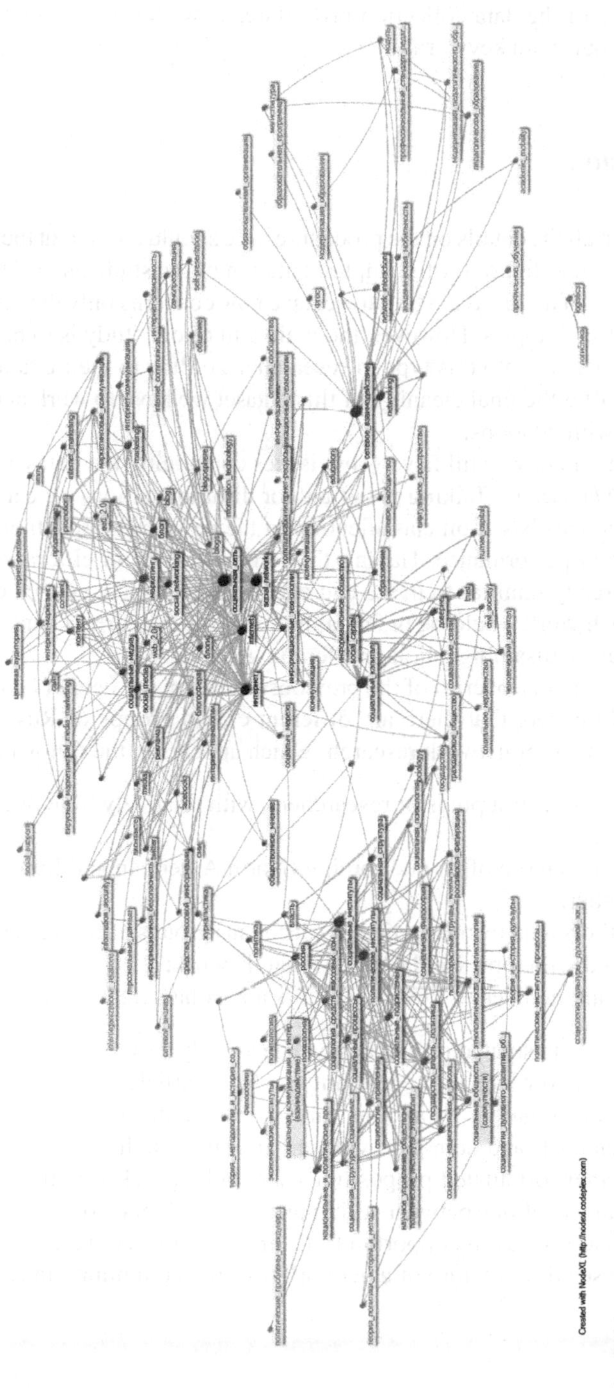

Fig. 3 Network of keywords

network topics. Another outcome that can be done after examining this network is the correctness of the data collection procedure, as we have not met any irrelevant topics among our main keywords.

5 Discussion

Having written all the details on the procedures, we should admit that there are a lot of things to be done in future for the implementation of the study and accomplishment of its aims. First of all, the dataset that we have now concerns only the information on articles on network topics. However, the main aim of our study is to build networks of scientists involved into this type of research according to their citation practices. It means that after the final cleaning of this dataset we have to work at collection of other dataset with citations.

At the same time, we still have some issues concerning the dataset with articles information (Dataset 1). Talking about author disambiguation, we can propose the usage of techniques based on classification of the network of coauthors, which can increase method performance. Having the information from eLibrary resource on the authors already familiar to it, we can check the efficiency of such classification technique. Such work would be practical not only for the current study, but also for solving the author disambiguation problem in general.

Basing on main discoveries of the previous studies in the field of citation analysis [30], we could expect that there are different closed groups of Russian scientists working in the field of network research, which appears in the following aspects:

- There are discrete groupings of researchers, with relatively little overlap between Russian authors;
- Russian authors more often cite foreign (North American and European) authors than each other;
- Russian authors tend to cite particular (different) groups of foreign authors, which are connected with topics and methods that they use;
- The significant number of Russian authors are isolated researches.

The last issue on isolated researches can be already verified with our preliminary results, according to which the significant number of articles in our collected sample does not have any citations from other authors, which means that they are invisible to the other scientists and do not provide any information that can be used by others in the field. Other formulated propositions should be checked in the future studies, during the analysis of ties between citing and cited authors. Analysis of the full set of data will also give us the opportunity to find the most active drivers of network studies and to see the structure of a network research community in Russia.

6 Conclusion

The main aim of this article was to present some methodological issues concerning our proceeding research on the network studies field in Russia. Providing the overview of the method of citation (reference) analysis as a tool for studying scientific fields, we showed its power and relevance to the studies of scientific communities structure. Then we described the process of data collection in deep details, in order anyone interested could repeat the data collection procedure. We emphasized some methodological and technical issues typical for the process of data collection, network expansion and filtering strategies, authors disambiguation and transliteration for the specific problem of domain-oriented information retrieval. Data collection and extraction code was published online,[3] so that any researcher could make experiments in his domain. We tried to enumerate all the problems that we faced to in our study and proposed the procedures of their overcoming. From one side these problems are standard for the data collection, but we also see some specific characteristics of the eLibrary resource. Providing the brief information on collected data on articles, we made a description of our dataset. We considered the thematic map of the data, which was constructed by network analysis of the main keywords used in the collected articles. Finally, basing on previous studies and collected data, we made some propositions that should be checked during next steps of analysis. These steps should be done in the following directions:

- on methodology: future work on the author disambiguation techniques based on classification of the network of coauthors, which can increase method performance;
- on data collection: collection of the full dataset on ties between authors (Dataset 2);
- on data analysis: the analysis of ties data base, which will allow us to build the network of research community of scientists involved into network studies in Russia and answer the main research questions of this study.

Acknowledgements The study has been funded by the Russian Academic Excellence Project '5–100' and RFBR grant 16-29-09583 "Methodology, techniques and tools of recognition and counteraction to organized information campaigns on the Internet". We thank the participants of the Sixth International Conference on Network Analysis NET 2016 and its organizer Laboratory of Algorithms and Technologies for Networks Analysis (LATNA) of National Research University Higher School of Economics (Nizhny Novgorod, Russia) for fruitful discussions and valuable comments on the platform of LATNA Laboratory.

[3] Available at https://github.com/lab533/elibrary.

References

1. Barabsi, A.L., Jeong, H., Neda, Z., Ravasz, E., Schubert, A., Vicsek, T.: Evolution of the social network of scientific collaborations. Phys. A **311**(34), 590–614 (2002)
2. Bar-Ilan, J.: Informetrics at the beginning of the 21st century-a review. J. Inf. **2**, 152 (2008)
3. Belotti, E.: Getting funded: multi-level network of physicists in Italy. Soc. Netw. **34**(2), 215–229 (2012)
4. Casey, D.L., McMillan, G.S.: Identifying the "Invisible Colleges" of the "Industrial and Labor Relations Review": a bibliometric approach. Ind. Lab. Relat. Rev. **62**(1), 126–132 (2008)
5. Cronin, B., Atkins, H.: The Web of knowledge-a Festschrift in honor of Eugene Garfield. ASIS Monograph Series. Information Today, Medford (2000)
6. De Bellis, N.: Bibliometrics and Citation Analysis: From the Science Citation Index to Cybermetrics. The Scarecrow Press, 417 p. (2009)
7. Devyatko, I.: Sotsiologicheskie teorii deyatel'nosti i prakticheskoy ratsional'nosti [Sociological theories of activity and practical rationality]. Chistie vodi, 336 pp (2003)
8. Farkas, I., Derenyi, I., Jeong, H., Nda, Z., Oltvai, Z.N., Ravasz, E.: Networks in lifescaling properties and eigenvalue spectra. Phys. A **314**(14), 25–34 (2002)
9. Garfield, E.: Citation Analysis as a Tool in Journal Evaluation. Science New Series **178**(4060), 471–479 (1972)
10. Gradoselskaya, G.: Setevye izmereniya v sotsiologii: Uchebnoe posobie [Network Measurement in Sociology: the textbook], Batygin G.S. (ed.). Publishing House New Textbook, 248 pp (2004)
11. Groenewegen, P., Hellsten, L., Leydesdorff, L.: Social networks as a looking glass on the social networks community. In: Sunbelt XXXV International Sunbelt Social Network, Abstracts. Hilton Metropole, Brighton, UK, 23–28 June 2015, p. 118 (2015)
12. Gubanov, D., Novikov, D., Chkhartishvili, A.: Sotsial'nye seti: modeli informatsionnogo vliyaniya, upravleniya i protivoborstva [Social networks: Information models of influence, control and confrontation]. Izdatel'skaya firma "Fiziko-matematicheskaya literatura" M., 228 pp (2010)
13. Hersovici, M., Jacovi, M., Maarek, Y. S., Pelleg, D., Shtalhaim, M., Ur, S. 1998. The shark-search algorithm. An application: tailored Web site mapping. Comput. Netw. ISDN Syst. **30**, 317–326 (1998)
14. Hoffman, K., Doucette, L.: A Review of Citation Analysis Methodologies for Collection Management. College and Research Libraries (2012)
15. Hummon, N., Carley, K.: Social networks as normal science. Soc. Netw. **15**, 71–106 (1993)
16. Kaufman, L., Rousseeuw, P.J.: Finding Groups in Data: An Introduction to Cluster Analysis. Wiley, 368 p. (2005)
17. Kharhordin, O.: Predislovie k Latur B. Nauka v deystvii: sleduya za uchenymi i inzhenerami vnutri obshchestva [Preface to B. Latour, Science in Action: How to Follow Scientists and Engineers through Society] (trans: St.Peter, F.K.). European University in St. Petersburg, pp. 7–19 (2013)
18. Kravchenko, S.: Sotsiologiya: paradigmy cherez prizmu sotsiologicheskogo voobrazheniya: uchebnik [Sociology: paradigm through the prism of sociological imagination: the textbook], 3rd ed. Examen, 750 pp (2007)
19. Levenshtein, V.I.: Binary codes capable of correcting deletions, insertions, and reversals. Sov. Phys. Dokl. **10**(8), 707–710 (1966)
20. Lewis, K.: Collaboration and acknowledgment in a scientific community. In: Sunbelt XXXV International Sunbelt Social Network, Abstracts. Hilton Metropole, Brighton, UK, 23–28 June 2015, p. 176 (2015)
21. Leydesdorff, L.: Theories of citations? Scientometrics **43**(1), 5–25 (1998)
22. MacRoberts, M.H., MacRoberts, B.R.: Problems of citation analysis: a critical review. J. Am. Soc. Inf. Sci. (1986–1998) **40**(5) (1989)
23. Moed, H.F.: Citation Analysis in Research Evaluation. Springer, Dortrecht (2005)

24. Newman, M.E.J.: The structure of scientific collaboration networks. Proc. Nat. Acad. Sci. U.S.A. **98**(2), 404409 (2001)
25. Otte, E., Rousseau, R.: Social network analysis. J. Inf. Sci. **28**, 441 (2002)
26. Radaev, V.: Osnovnye napravleniya razvitiya sovremennoy ekonomicheskoy sotsiologii [The main directions of development of modern economic sociology]. In: The book Ekonomicheskaya sotsiologiya: Novye podkhody k institutsional'nomu i setevomu analizu [Economic Sociology: New approaches to institutional and network analysis] (trans: Dobryakova M.S.); M.: POSSPEN, pp. 3–18 (2002)
27. Safonova, M., Winer, B.: Setevoy analiz sotsitirovaniy etnologicheskikh publikatsiy v rossiyskikh periodicheskikh izdaniyakh: predvaritel'nye rezul'taty [Network analysis of co-citations of ethnological publications in Russian periodicals: preliminary results]. Sotsiologiya: Metodologiya, metody. matematicheskoe modelirovanie **36**, 140–176 (2013)
28. Smith, L.: Citation analysis. Library Trends (1981)
29. Sokolov, M., Safonova, M., Guba, K., Dimke, D. 2012. Intellektual'nyy landshaft i sotsial'naya struktura lokal'nogo akademicheskogo soobshchestva (sluchay peterburgskoy sotsiologii) [Intelligent landscape and social structure of the local academic community (the case of the St. Petersburg Sociology)]. M.: Izd. dom Vysshey shkoly ekonomiki, 48 pp
30. Tight, M. Citation and co-citation analysis. In: Brew, A., Lucas, L. (eds.) Academic Research And Researchers: Policy and Practice. Open University Press (2009)
31. Vakhshtain, V.: Vozvrashchenie material'nogo. Prostranstva, seti, potoki v aktorno-setevoy teorii [The return of the material. "Spaces", "networks", "flows" in the actor-network theory]. Sotsiologicheskoe obozrenie **4**(1), 94–115 (2005)
32. Wagner, C.S., Leydesdorff, L.: Network structure, self-organization, and the growth of international collaboration in science. Res. Policy **34**(10), 1608–1618 (2005)
33. Ward, J.H.: Hierarchical grouping to optimize an objective function. J. Am. Stat. Assoc. **58**(301), 236–244 (1963)
34. Warner, J.: A critical review of the application of citation studies to the research assessment exercise. J. Inf. Sci. **26**, 453 (2000)
35. White, H.D., Wellman, B., Nazer, N.: Does citation reflect social structure? longitudinal evidence from the Globenet interdisciplinary research group. J. Am. Soc. Inf. Sci. Technol. **55**(2), 111–126 (2004)
36. Winer, B., Divisenko, K.: Kognitivnaya struktura sovremennoy rossiyskoy sotsiologii po dannym zhurnal'nykh ssylok [Cognitive structure of modern Russian sociology according to the journal links]. Zhurnal sotsiologii i sotsial'noy antropologii **4**(60) (2012)
37. Zhang, T., Ramakrishnan, R., Livny, M.: BIRCH: an efficient data clustering method for very large databases. In: Proceedings of the 1996 ACM SIGMOD International Conference Management data SIGMOD 96, vol. 1, pp. 103–114 (1996)